Experimental Methodology for Human–Robot Interaction

Leading figures in Australian robotics research provide an overview and guidance for human-robot interaction (HRI) experimental design and evaluation methodologies that consider the ethical implications of the research and its applications from a human-centred and contextual perspective. The authors explain introductory and advanced topics in HRI with a focus on human-centred evaluation and ethical practices. They also provide an online interactive checklist tool for novice HRI researchers and students to deploy when designing their own studies.

The book is structured into three parts. In Part I, the authors first review fundamental methodologies and provide an interactive checklist tool of the HRI experimental study life cycle to guide beginners to the field. Part II introduces an expanded set of approaches to support researchers and practitioners to create high-quality study designs that draw on practices from human-computer interaction, human-centred artificial intelligence, psychology and social science, and advance ethical HRI research. Finally, in Part III, the authors discuss a selection of HRI studies as examples of how the introduced methodologies are adopted, which will support the readers to further understand the fundamental and advanced methodologies described in Parts I and II. The diverse collection of case studies enables readers to grasp the state of the art and apply what they have learned in their own practices.

This book is a vital resource for both students new to the field and experienced researchers and practitioners. The book's practical focus and clear elucidation of relevant case studies, from its introduction to the HRI experimental study life cycle through to advanced methods emerging in the field, ensures that this will greatly benefit progress in the field with human-centred and ethical experimental methodology.

Experimental Methodology for Human–Robot Interaction

Guidelines and Case Studies for Human-Centred and Ethical Robotics Research

Leimin Tian, Tina L. Y. Wu, Nicole L. Robinson, Pamela Carreno-Medrano, Wesley P. Chan, Maram Sakr, Elahe Abdi, Elizabeth A. Croft, and Dana Kulić

CRC Press
Taylor & Francis Group
Boca Raton London New York

CRC Press is an imprint of the
Taylor & Francis Group, an **informa** business

Designed cover image: *Cover image painted by Dr. Tina L.Y. Wu. Reprinted with permission from her family.*

First edition published 2025
by CRC Press
2385 NW Executive Center Drive, Suite 320, Boca Raton FL 33431

and by CRC Press
4 Park Square, Milton Park, Abingdon, Oxon, OX14 4RN

CRC Press is an imprint of Taylor & Francis Group, LLC

ISBN: 978-1-032-69361-3 (hbk)
ISBN: 978-1-032-69365-1 (pbk)
ISBN: 978-1-032-69368-2 (ebk)

DOI: 10.1201/9781032693682

Typeset in Latin Modern font
by KnowledgeWorks Global Ltd.

In loving memory of
Tina and Andrew.
Your adventure continues beyond.

Contents

Preface

In this book, we provide an overview and guidance for Human-Robot Interaction (HRI) experimental design and evaluation methodologies. We consider the ethical implications of the research and its applications from a human-centred and contextual perspective.

This book will benefit researchers and students new to the field with its introduction to the HRI experimental study life cycle. Further, it will benefit experienced researchers and practitioners by introducing advanced methods emerging in the field. The diverse collection of case studies will allow readers to grasp the state of the art and apply what they have learned in their own practices.

The first part of the book introduces the fundamental components of the life cycle of an HRI research study. It is aimed at researchers and students who are new to the field to get an overview of the basics in designing, conducting, and reporting an HRI study. Using an interactive checklist, the readers are provided with a systematic tool to develop their own HRI studies. We also review key methodologies that improve the validity, repeatability, and generalisability of the study outcomes, and provide guidance for the design of new HRI systems that are effective based on scientific evidence.

The second part of the book introduces an expanded set of approaches to support experienced HRI researchers and practitioners to create high-quality study designs that advance ethical HRI research and applications. We provide guidance for emerging HRI study design and evaluation methodologies that consider the ethical implications of the research and its applications from a human-centred and contextual perspective. We also draw from interdisciplinary practices in these advanced topics of HRI.

The third part of the book investigates a collection of example HRI studies with diverse user populations, application areas, and robotic platforms to demonstrate how the HRI research methodologies presented in Parts I and II are utilised in state-of-the-art research. We also discuss the challenges and future directions of HRI summarised from the methodologies and case studies discussed in this book.

Our hope is that readers will come away with a systematic approach to the HRI experimental study life cycle, and benefit from the ethics and interdisciplinary centred approach to HRI study design presented in Parts II and III.

Author Bios

Leimin Tian is a senior research scientist and the leader of the Human-Robot Interaction team at CSIRO Robotics. Her research spans over human-robot interaction, human-centred AI, and affective computing. She focuses on developing robots that incorporate social-emotional intelligence to enhance communication and collaboration in various applications. She attained her BEng from Beijing University of Posts and Telecommunications in 2012 and her MSc and PhD from The University of Edinburgh in 2013 and 2018, respectively.

Tina L. Y. Wu (1996–2023) obtained her PhD degree from Monash University in 2024. She obtained her BASc in Biomedical Engineering (Honours) from the University of Waterloo in 2019. Her research focused on human-robot interaction, human motion analysis, rehabilitation robots, and user experience design.

Nicole L. Robinson is a lecturer at Monash University. She is the Co-founder, Executive Director, and Board Member of LYRO Robotics Pty Ltd and is an Advisor to the Australian Government's National Artificial Intelligence Centre hosted by CSIRO. She is also an advisor to Monash Robogals. She attained her PhD from Queensland University of Technology in 2018.

Pamela Carreno-Medrano is a lecturer at Monash Robotics. Her research interests lie at the intersection of human-robot interaction, human-centred and embodied AI, and human-centred design. Her work focuses on developing autonomous, intelligent agents (virtually and/or physically embodied) that can adequately interact with humans in both short- and long-term scenarios. She obtained her bachelor's degree from Universidad EAFIT, Medellin, Columbia, in 2010; her master's degree from École Nationale d'Ingénieurs de Brest in 2012; and her PhD from Université de Bretagne Sud in 2016.

Wesley P. Chan is a Senior Software Engineer at A&K Robotics and an Affiliate of the Department of Electrical and Computer Systems Engineering at Monash University. He attained his bachelor's and master's degrees from the University of British Columbia in 2010 and 2012, respectively, and his PhD from the University of Tokyo in 2016.

Maram Sakr is a PhD candidate in the joint program between the Department of Mechanical Engineering at the University of British Columbia, Canada, and the Department of Electrical and Computer Systems Engineering at Monash University, Australia. She is an experimental roboticist working at the intersection of robotics, machine learning, and Human-Robot Interaction. Her research focuses on enabling everyday users to collaborate with and teach robots intuitively and efficiently.

Elahe Abdi is a Senior Lecturer of Robotics at Monash University. She obtained her BSc from the University of Tehran in 2009, her MSc from Sharif University of Technology in 2012, and her PhD from the Swiss Federal Institute of Technology Lausanne in 2017.

Elizabeth A. Croft is Vice President Academic and Provost, University of Victoria, Canada. She obtained her BASc from the University of British Columbia in 1988, MASc from the University of Waterloo in 1992, and PhD from the University of Toronto in 1995.

Dana Kulić is a leading expert in robotics, specialising in human-robot interaction and intelligent systems. Her work focuses on developing autonomous systems that interact naturally with humans and learn from user feedback. She is a Professor at Monash University and is the founding Director of Monash Robotics. She attained her BASc, MASc, and PhD from the University of British Columbia in 1998, 1998, and 2006, respectively.

Chapter Authors

Aimee K. Allen is a PhD student at Monash Robotics. She authored Chapter 16 in Part III.

Zhongxiang Chen is a PhD student at Monash Robotics. He authored Chapter 17 in Part III.

Yanjun Yang is a research fellow at Monash Institute of Railway Technology, Faculty of Engineering, Monash University. He obtained his PhD from Monash Robotics in 2023. He authored Chapter 14 in Part III.

Acknowledgements

The authors would like to thank Prof. Sharon Oviatt and Prof. Robert Sparrow for their feedback on earlier drafts of the manuscript.

Part I

Introduction to Human-Robot Interaction (HRI) Experiment Methodology

1

Life Cycle of an HRI Study and an Interactive Study Design Checklist

This chapter introduces the field of experimental human-robot interaction (HRI) and provides a model for the common life cycle of an HRI experiment study. To help practitioners design and execute their own experiments, we also introduce an interactive checklist for designing, conducting, and reporting an HRI experiment study: `https://doi.org/10.26180/26507314`.

1.1 Introduction

Huma-Robot Interaction (HRI) is a fast-growing discipline in robotics that focuses on how robots and humans interact or collaborate. It adopts human-centred design and research approaches [Shneiderman (2022)] that evaluate how a robot system may address the needs of an individual or a population, its benefits as perceived by these users, and the collaborative success of the human and the robot [Robinson et al. (2022)].

Experiments involving human participants are commonly employed to validate proposed approaches and theoretical frameworks. Similar to experimental research in other disciplines, the typical life cycle of an HRI empirical study can be divided into four phases: Concept and experiment design (e.g., identifying hypotheses), experiment execution (e.g., participant recruitment), analysis (e.g., data visualisation), and post-study phase (e.g., open-sourcing code). A detailed breakdown of the HRI study life cycle will be discussed in Section 1.2.

As a result of focusing on human and contextual factors beyond robotic technologies, HRI research requires a multidisciplinary theoretical and experimental framework, adopting methodologies from human factors, Human-Computer Interaction, psychology, ergonomics, and other complementary fields. For example, Prati et al. (2021) adopted tools from user experience research, namely, user analysis, task analysis, user/task matrix, and experience maps, in evaluating human-robot collaborative

DOI: 10.1201/9781032693682-2

assembly tasks in an industrial environment. As will be further discussed in Chapter 9, interdisciplinary approaches in HRI research lead to opportunities for expanding experimental methodologies, while simultaneously presenting additional challenges.

As an emergent field, HRI research often employs exploratory studies with less systematic methodology; this is considered a common and reasonable approach for early-stage scientific research [Ledgerwood et al. (2017), Shrout and Rodgers (2018)]. Most existing studies in HRI have provided an initial prototype to explore initial concepts and outcomes, for example, an uncontrolled pilot with a small sample size on one key measurement [Hoffman and Zhao (2020)]. As the HRI field is becoming more well-established, it is vital to employ more systematic experimental methodologies to establish strong principles and answer more realistic research questions about design, implementation, and deployment. As identified by Jung and Hinds (2018), understanding the interaction context and a robot's influence on people at the individual, group, and societal levels beyond lab-based studies, that is, in-the-wild evaluation, is key to developing robust HRI theories. At the same time, conducting a systematic analysis of a factor in HRI may involve reducing the complexity and realism of the study design to gain more control and isolate the factor of interest. It is often challenging to balance realism and rigour in HRI research. Further, to ensure beneficial outcomes, ethical guidelines for designing and conducting HRI studies are required [Axelsson et al. (2024)], which we will further discuss in Chapter 8.

Now we will discuss the remainder of this chapter. In Section 1.2, we provide an overview of the life cycle of a typical HRI study, which guides the discussion of methodological guidelines in this book. We then briefly review existing guidelines for HRI experiments in Section 1.3 to provide the background and motivation for this book. Finally, in Section 1.4 we summarise shared recommendations in these existing guidelines and their connections with chapters in this book. Common terminologies used in this book are given as follows:

- *HRI system/application*: A physical or virtual robotic system or application that achieves intended outcomes by interacting with people, for instance, a service robot that takes and delivers orders to patrons at a restaurant.

- *HRI study/experiment/research*: A research study that conducts experiments to evaluate specific aspects of HRI, for example, a lab study evaluating how a robot with the ability to express emotions influences a participant's behaviours during a collaborative task. An HRI study can inform the design and expected outcomes of a specific HRI system or provide a more general understanding of humans' mental states or behaviours in HRI.

- *Stakeholders of an HRI system*: They are the primary and secondary users of an HRI system that the system can directly or indirectly influence, for instance, patrons and staff in a restaurant employing a service robot. Note that there are other stakeholders of an HRI system in the broader application context, such as researchers, designers/developers, policy makers, and regulators, which we will discuss further in Chapter 8.
- *Participants in an HRI study*: People recruited to participate in an HRI study to infer how specific aspects of an HRI system may be expected or received by stakeholders, for example, those performing the collaborative task with the robot in the example above.

1.2 Life Cycle of an HRI Study

A systematic description of the HRI study life cycle, which captures the whole process from conceiving a research question to post-study follow-ups, can guide researchers to follow best methodological practice to increase scientific rigour and help avoid a possible HRI replication crisis. For instance, Hoffman and Zhao (2020) presented a primer for conducting HRI experiments following nine chronologically ordered stages of empirical research activities. This systematic description was aimed to provide a baseline for researchers new to the field, thus helping them avoid common mistakes and encourage more reproducible studies.

While individual HRI studies differ in the exact activities taken, the life cycle of an HRI study can often follow a similar pattern, even when different robots, metrics, or environments are the key focus of the research. We therefore propose a model of the life cycle of a typical HRI study as shown in Figure 1.1.

As shown in our proposed life cycle, an HRI study can be divided into four stages, namely, the design phase, the experiment phase, the analysis phase, and the post-study phase. Note that the design phase is iterative, as indicated by the arrow linking from the pilot step back to the research question step. It is common in HRI research to take several iterations through the design stage at progressive levels of detail and rigour. This iterative design process ensures that the research questions and hypotheses are refined and the experimental methods are aligned to investigate them. Further, Figure 1.1 is illustrative of the investment required to yield a successful study outcome, with the design phase being the most time-consuming and needing the most effort to achieve suitable experimental protocols. Without a well-refined design phase, the experiment phase can suffer costly failures or introduce problematic data to the analysis phase. In addition, the life cycle does not terminate when an HRI study has concluded.

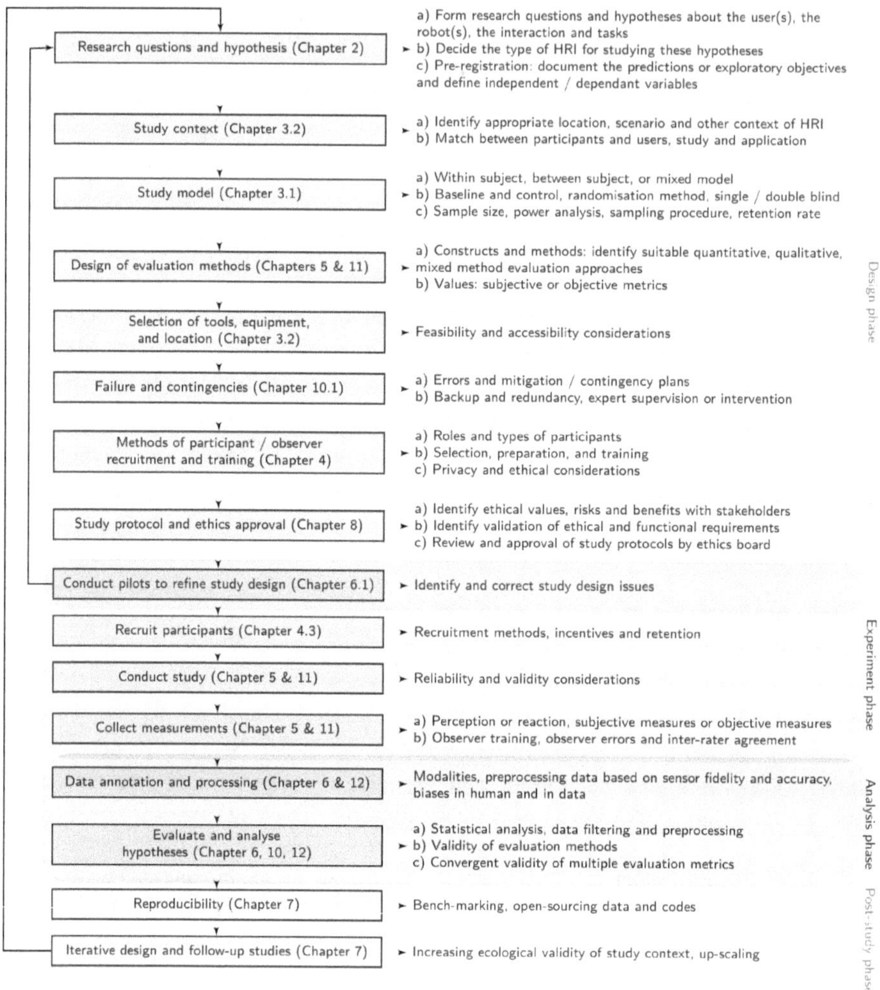

FIGURE 1.1
Life cycle of an HRI study. We also provide an interactive tool that helps readers follow the discussion in this book and design their own study: `https://doi.org/10.26180/26507314`.

As shown by the arrow linking back to the start of the life cycle in the post-study phase, transparent reporting of the study design and results can inspire follow-up studies that continue to progress to HRI research.

Each step in the HRI study life cycle will be explained in further detail in relevant chapters in this book, as linked in Figure 1.1. In these chapters, we discuss specific recommendations on experimental approaches. We also illustrate how the proposed recommendations can be applied in practice with examples in existing HRI literature. As HRI is an interdisciplinary research field, we also include methodological approaches from complementary fields, including Human-Computer Interaction, human-centred Artificial Intelligence (AI), psychology, and cognitive science.

An interactive tool accompanying our discussion of HRI experimental methodologies can be accessed at `https://doi.org/10.26180/26507314`. This web-based tool can be opened in a browser with all usage data stored locally. This tool directs users to brief descriptions of each step in the HRI study life cycle as shown in Figure 1.1. It is also an interactive checklist that allows users to enter and save details regarding their own study while following the discussion in this book.

1.3 Background on HRI Experimental Research

In this section, we provide a brief overview of guidelines for HRI experimental research to motivate the objective of this book, namely, to offer practical guidelines and to introduce an expanded set of methodologies that facilitate the design and evaluation of HRI systems.

1.3.1 Need for an HRI Study Methodology Guideline

HRI research has an urgent need to work towards a scientifically rigorous methodology that ensures the validity and generalisability of the findings. A review by Robinson et al. (2019) reveals that HRI studies are often designed and executed based on the preferences of researchers and available resources: In relation to research design, less than 10% of studies for social robot applications in healthcare have deployed randomised controlled trials, a research methodology that can address issues regarding methodological rigour and researcher bias when concluding the effectiveness of social robots for a use case domain. In the elderly care domain, two-thirds of studies had fewer than 30 participants, while the majority lasted a day, and only 11% lasted 30 days or more. Most importantly, 90% of the studies lacked reference to an established theoretical framework [Zafrani and Nimrod (2019)].

Recently, increased attention and importance have been placed on adopting rigorous study methodology across multiple scientific fields. For example, behavioural and social science is experiencing a "replication crisis" [Maxwell et al. (2015)], in which efforts to replicate existing studies often do not yield the same observations or conclusions. A lack of replication can impact the validity of the proposed findings, lead to incorrect assumptions about the studied phenomena for future work, and create resource waste to revise or verify early work. In recent years, there has been a growing effort to review the methodology in existing HRI studies. These reviews are important for the research community to improve the scientific outcomes of experimental HRI work, identify limitations of current approaches, and recommend better practices. Existing reviews in the domain of HRI have mainly summarised current practices, rather than providing a guideline for future studies.

There are existing guidelines for conducting HRI studies aimed to introduce new researchers to the study process, as we will further discuss in Section 1.3.2. To aid the review of HRI study methodologies, there have also been systematic approaches to categorising HRI studies, as will be discussed in Section 1.3.3. In addition, a number of reviews have focused on providing detailed guidelines for one key step during the HRI study life cycle, a single methodology of HRI research practice, or one type of assessment approach, as we will discuss in Sections 1.3.4 and 1.3.5. Further, there have been reviews of HRI research for a specific application domain or user population, which we further discuss in Section 1.3.6.

These representative examples of existing reviews and guidelines for HRI methodology motivate the proposed HRI life cycle to provide clear links between single discipline or domain reviews into a more cohesive picture. The HRI guidelines proposed in this book will draw from different HRI studies to present a detailed and holistic view, while maintaining a focus on recommending suggestions applicable to each of the core steps during the study life cycle.

1.3.2 Tutorials for Conducting HRI Studies

Bartneck et al. (2020b) provide a comprehensive introduction to HRI studies. This report covered good practices in HRI studies, namely, research question (explanatory vs. confirmatory), study design (qualitative vs. quantitative vs. mixed methods), participant (choice of population and sample size), context (location, time, and social unit), robot appearance and functionality, mode of interaction (teleoperation vs. physical vs. simulated), as well as direct and indirect metrics, statistical analysis, and ethical considerations. Their discussion of the (dis)advantages of each approach and analysis of the best practices provides a comprehensive introduction to the HRI study. Similarly, Hoffman and Zhao (2020) proposed an introduction to HRI methodology and a tutorial for graduate students and researchers new to the field. A hypothetical HRI study was used as an example to illustrate the process of how an HRI study is conceived and conducted. In addition, Jost et al. (2020) provided a much needed interdisciplinary perspective on HRI evaluations. The authors focused on HRI methodologies that are human-centred and contextual and reviewed qualitative and mixed-method approaches with examples in HRI.

1.3.3 Categorisation of HRI Studies

HRI studies have been categorised in multiple review papers based on a number of attributes. Various taxonomies have been proposed in the literature for classifying HRI. These include taxonomies classifying HRI by the type of study [Damacharla et al. (2018)], the role of the robot [Collins (2019)], its morphology [Yanco and Drury (2004)], level of autonomy [Thrun (2004), Beer et al. (2014)], behaviour [Shim and Arkin (2013)], and characteristics [Rahimi and Karwowski (1990)], the role of the human [Scholtz (2003)], type of task [Yanco and Drury (2004), Steinfeld et al. (2006)] and interaction [Seibt (2017), Olsen and Goodrich (2003)], as well as human-robot proximity [Goodrich and Schultz (2007), Agah (2000)], and coordination [Jiang and Arkin (2015), Hoffman (2019b)]. The HRI study themes, types, and challenges, as of the year 2008, were reviewed by Goodrich and Schultz (2007). Five attributes of HRI were highlighted including autonomy, task type, and structure of the human-robot team. In a similar approach, Baxter et al. (2016) categorised papers accepted at the ACM/IEEE International Conference on Human-Robot Interaction between years 2013–2015 based on the level of robot autonomy, participant populations, evaluation environments, length of empirical studies, approach to statistics, and reproducibility. Based on their review, Baxter et al. (2016) provided a set of recommendations, highlighting the importance of clarity of the research goal, level of robot autonomy, and statistical analysis, as well as justification of the number of participants, length of interaction, and reproducibility of the study. The establishment of standards and common metrics has been reported as an emerging effort in these reviews, with the HRI community still progressing towards this goal a decade later.

1.3.4 Hypotheses Based on Human-Human Interaction Theories

Hypotheses in HRI studies are often grounded by theories from other relevant disciplines, such as psychology and cognitive science. Thus, it is inevitable that the recent replication crisis in psychology has also influenced the field of HRI. For example, Irfan et al. (2018) presented their failures to replicate the social facilitation theory, that is, the presence of others resulting in a positive influence on a person's task performance, in HRI context. Based on this experience, they recommended that HRI researchers who are consumers of psychology literature avoid older psychology literature with weak methods and encouraged bench-marking efforts in HRI. Their work highlighted the challenges of observing and measuring social interaction in HRI. Similarly, Leichtmann and Nitsch (2020) reviewed HRI studies on human-robot physical and social distance and investigated whether it is appropriate to generate hypotheses for HRI studies based on human-human interaction (HHI) theories. They found that HRI study hypotheses are often derived from HHI under the assumption that robots are treated as social actors in a manner similar to humans. However, this assumption may not always be valid. Based on their review, the authors recommended pre-registration of hypotheses, methodology, and analysis protocol, as well as transparent reporting of HRI studies grounded with HHI theories. In Chapter 2, we will further discuss considerations on developing research questions and hypotheses for HRI studies.

1.3.5 Focused Reviews of HRI Study Components

Existing reviews of HRI methodology often focus on one particular aspect or evaluation method. For example, Lasota et al. (2017) reviewed factors influencing safety in HRI and different measurements of safety; Young et al. (2011) reviewed the use of biological features and means to incorporate user variances; and Prewett et al. (2009) reviewed the measurement of robot teleoperator workload in relation to task outcomes. It is important to have such in-depth investigations focusing on one type of HRI or on one aspect of HRI evaluation. However, identifying a set of guidelines that inform on core steps of an HRI study cycle is crucial to improve the overall quality of HRI studies.

The majority of HRI methodology reviews have focused on the evaluation aspect and provided recommendations regarding the design phase of the HRI study life cycle. As an example, focusing on the participant selection and recruitment step, Lee et al. (2022) reviewed papers published at the Association for Computing Machinery (ACM) HRI conference between 2006 and 2021 in terms of human roles and population types in these studies. Their review indicated that there is a need for more diverse and representative participants in HRI studies. In another example, Bethel and Murphy (2010) identified study size and lack of multiple assessment methods as the primary issues in existing studies. They argued that a large sample size better represents the target population and exhibits a higher probability for statistically significant results. They also suggested that the deployment of three or more evaluation metrics for the same construct produces more reliable and accurate results.

Many metrics used in HRI studies remain "observed" with a lack of functional or generalisable measurement mechanisms. For example, Murphy and Schreckenghost (2013) proposed an HRI metric taxonomy based on a review of 29 studies. A total of 42 metrics were categorised into three groups: human (e.g., reliability, productivity, and awareness), robot (e.g., plan state and time in manual/autonomous operation), and system (e.g., efficiency, safety, and co-activity). Psychological assessment metrics such as cardiovascular, electro-dermal, and brain activity are increasingly used to evaluate human responses in HRI [Tiberio et al. (2013)]. When studying social interactions and human responses in HRI, the most commonly adopted approach is to use the Likert scale to measure subjective judgements of participants. However, Schrum et al. (2020) reviewed the use of the Likert scale in HRI studies and found that only 3 out of the 110 papers performed their analyses correctly. This demonstrates the urgent need for methodological guidelines to ensure the quality and robustness of HRI studies, especially regarding the validity of evaluation methods and appropriate analyses of different metrics. Another example of a review on specific HRI methodology is Weiss and Spiel (2022), which focused on participatory approaches to HRI and discussed robot-specific challenges, namely, robot narratives influenced by existing expectations and the portrayal of robots in the mass media. Others challenges included hardware constraints of robots and adaptivity and unpredictability of machine learning models used by robots.

In contrast to reviews focusing on the design phase of HRI studies, Belhassein et al. (2019) provided recommendations for the experiment and post-study phases of HRI user studies with a focus on participants and the replication crisis. Regarding the experiment phase, they suggested that the outcome of the HRI study can be improved by recruiting more users with diverse backgrounds, rigorous implementation of the protocol, allowing time for the user to get accustomed to the robot, ensuring physical and psychological safety, assessing the suitability of the robot for the task, choosing the right and preferably standardised measurements, deployment of theoretically solid tools and measures, and finally providing the required details for reproducing the study. Regarding the post-study phase, they suggested standardisation of the technical components to address study reproducibility.

1.3.6 Real-Life HRI Scenarios and Technical Limitations

Many HRI systems are designed with the goal of application to real-life education, health, and elderly care scenarios [Dautenhahn (2007)]. To assess the outcomes of such HRI systems, it is critical to evaluate HRI in-situ. For example, Weiss et al. (2009) proposed the usability, social acceptance, user experience, and societal impact (USUS) framework. Realistic HRI relies on advancements in sensory technologies and algorithms in robotics [Thomaz et al. (2016)]. To envision new research questions beyond existing technical restraints, especially when studying complex social interactions, the Wizard of Oz (WoZ) approach is often applied in HRI studies, where a hidden human operator controls the robot remotely while the participants were under the impression that the robot was fully autonomous. The WoZ approach facilitates experiment safety and control of variable factors in HRI. Riek (2012) reviewed how WoZ has been used in HRI to identify valid designs. They stressed the importance of operator training and instruction, as well as the formation of hypotheses on user and robot behaviours. In addition to WoZ, HRI studies often involve simulated interaction scenarios and experiments in controlled lab environments to isolate the factor of interest. While these are necessary first steps, in-the-wild HRI studies with target users are attracting growing attention in the field [Jung and Hinds (2018)]. In Chapter 3, we will further discuss considerations on HRI study contexts.

1.4 Summary

This chapter introduced the typical life cycle of an HRI experimental study, as shown in Figure 1.1. This overview guides the discussion in Parts I and II of this book on methodological guidelines at each key step of the life cycle. Using the open-source interactive tool we developed (`https://doi.org/10.26180/26507314`), a researcher may follow the discussion and create a custom checklist of their own HRI study.

Our review of existing surveys of HRI study methodologies motivates the need for an extended set of guidelines to key steps in the full HRI study life cycle, as presented in Parts I and II of this book. For HRI researchers new to the field, this book offers an overview of the HRI study life cycle and introduces them to good practices for conducting HRI studies. For experienced HRI researchers, we provide in-depth analysis of the strengths and weaknesses of existing HRI methodological approaches, as well as insights on novel methods that can be adopted in future studies. Further, Part III includes case studies of experimental methodologies used in example HRI studies that put the guidelines into practice.

Because of the wide variation in research questions and experimental settings in HRI, a one-size-fits-all methodological approach across all HRI studies is not appropriate. However, we have identified five shared recommendations emerging from existing reviews of HRI studies:

Shared recommendations emerging from existing HRI reviews

1. *Relevance of psychology and social science theories to HRI*: Grounding HRI hypotheses with theories in psychology and social science helps to increase the strength of the HRI hypotheses, especially when the focus is on human perception and reaction during HRI [e.g., Groom et al. (2010)]. However, there remain significant differences between human-human interaction and HRI, which can limit the applicability of psychology or social science theories in HRI. This will be discussed further in Chapters 2 and 9.
2. *The importance of context*: HRI is context dependant. Researchers are encouraged to have an interdisciplinary perspective when conducting HRI studies and to incorporate social and interactive context in their research, such as the target user population or the location of the interaction. This will be discussed further in Chapters 3 and 4.
3. *Increase validity of evaluation by using multiple metrics*: The validity of evaluation metrics used is directly related to the validity of an HRI study. Researchers are recommended to use multiple metrics and evaluation approaches to measure a parameter and to investigate the convergence of these metrics. In addition, researchers are encouraged to use appropriate statistical tests and conduct power analyses when studying experimental results. This will be discussed further in Chapters 5, 6, 10, 11, and 12.
4. *Ecological validity of a study*: Researchers are encouraged to conduct field studies and longitudinal HRI studies. In HRI experiments, it is recommended to recruit participants from the target populations and pay attention to sample size when analysing the results. This will be discussed further in Chapters 3 and 4.
5. *Rigour and transparency are key*: To improve the reproducibility and generalisability of HRI studies, researchers are encouraged to pre-register their research questions and hypotheses. It is necessary to cover both significant and non-significant results in the discussion and open-source the code and data used in the work if possible. This will be discussed further in Chapters 2, 7, 8, and 10.

2

Developing Research Questions and Hypotheses

This chapter introduces exploratory and confirmatory HRI studies. We will discuss how to identify good research questions and corresponding hypotheses, which mark the beginning of the HRI study life cycle and the experimental design process.

2.1 From Research Question to Hypothesis

2.1.1 Exploratory and Confirmatory Studies

The experimental design process begins with identifying and refining the research question(s) the experimenter wishes to explore. An effective method for developing strong research questions begins with a problem statement describing a knowledge gap. The identification of this gap can be developed through literature review [Bartneck et al. (2020a)], personal observation, drawing similarities to other fields [Collins (2019)], or exploratory/pilot studies, among other methods. Once a research question is defined, a hypothesis – a testable claim – can be developed to answer the research question.

Current HRI studies often take the form of an exploratory or confirmatory study or a combination of both [Bartneck et al. (2020b)]. Exploratory or hypothesis-generating studies aim to formulate a new idea or gather a preliminary understanding of it when prior knowledge is limited. For example, in a study exploring how humans would interact with robots when robots begin to enter people's homes in the future, the investigators designed five likely domestic scenarios and studied how participants responded to different robot behaviours in these scenarios [Koay et al. (2009)]. Recently, there has been a growing number of exploratory HRI studies adopting user-centred and scenario-based approaches to generate hypotheses through collaboration with the stakeholders [e.g., Rueben et al. (2021), Zamfirescu-Pereira et al. (2021a)]. Among them, Rueben et al.

DOI: 10.1201/9781032693682-3

(2021) explored how people form a mental model of a robot's capabilities. Based on their observations from a field deployment of the robot and their interviews with the participants, they generated several hypotheses that can be further investigated to understand primary users' and bystanders' perceptions and experiences during HRI outside of the laboratory. Exploratory studies can also be used to identify the ethical and functional requirements related to an HRI system and are especially useful for investigating emergent, complex, and dynamic phenomena in HRI [Fischer (2021), Zamfirescu-Pereira et al. (2021a)].

Confirmatory or hypothesis-testing studies aim to validate expectations on how factors of interest (e.g., robot's shape, system design, and participant gender) will impact specific aspects of HRI. Confirmatory studies focus on validating initial research designs and are important for understanding the applicability and robustness of features implemented to satisfy the functional and ethical requirements of an HRI system. For example, based on the knowledge that non-verbal communications play an important role in human-human collaborations, the experimenters in Breazeal et al. (2005) designed a study to confirm the importance of non-verbal communication in HRI.

Observations in confirmatory studies can, in turn, inspire new exploratory studies. Thus, exploratory and confirmatory HRI studies together form a cycle of proposing and validating new insights [Veling and McGinn (2021)]. Thus, HRI research combining exploratory and confirmatory studies is desired to facilitate sustainable scientific discoveries.

2.1.2 Hypotheses in Confirmatory Studies

Focusing on confirmatory HRI studies, a key initial step is to develop clear hypotheses about the user(s) and the robot(s). A hypothesis is a predictive statement on how a given condition will affect a certain aspect of the outcome. Because the hypothesis drives the selection of the type of study and the method of evaluation [Bethel and Murphy (2010)], it needs to identify key constructs and be testable and concise [Rind (2011)]. As an example, in an HRI study examining the effects of non-verbal communication on the efficiency in human-robot teamwork [Breazeal et al. (2005)], the experimenters hypothesised that:

> "implicit non-verbal communication positively impacts human-robot task performance with respect to understandability of the robot, efficiency of task performance, and robustness to errors that arise from miscommunication."

This statement relates the given condition of the communication method used to the outcome of task performance. It names key constructs, including implicit communication and robot understandability, making a concise prediction that, by using implicit non-verbal communication, robot understandability, task efficiency, and robustness to errors from miscommunication will be improved. Because the aim of confirmatory user studies is to find support for the formulated hypotheses, a good hypothesis is testable in a scientific experiment. The condition stated in the hypothesis involves a factor, often referred to as an independent variable, that the researcher can vary and control independently in an experiment. The outcome stated in the hypothesis, also known as the dependent variable, is a measurable or observable parameter in an experiment setup.

Considering the preceding example, the factor of robot communication method was controlled independently from other factors by the researchers through employing either explicit communication only or both explicit communication and implicit non-verbal behaviours. The outcome of task performance was measured through questionnaires asking users to rate the effectiveness of the interaction, as well as through video analysis. While formulating a hypothesis has the benefits of focusing the research and guiding the design of the experiment, it is equally important for researchers not to be biased by the formulated hypotheses, such that they become blinded to unexpected interesting findings [Rind (2011)].

2.2 Building Hypotheses Based on HRI Attributes

When formulating hypotheses, it is helpful to think from a designer's perspective, as suggested by Goodrich and Schultz (2007). They defined HRI as "a field of study dedicated to understanding, designing, and evaluating robotic systems for use by or with humans". In their paper, they listed five attributes that the designer can change to affect HRI:

1. Level and behaviour of autonomy
2. Nature of information exchange
3. Structure of the human-robot team
4. Adaptation, learning, and training of people and robots
5. Nature of the task

A common strategy for formulating hypotheses is then to consider which attribute is the main concern of the study. Once the main attribute of concern has been identified, the levels for testing this attribute can be determined. These attribute settings would then naturally become the independent variables in the hypotheses. The dependent variables stated in a hypothesis often relate to what the designer hopes to achieve.

In the following sections, we take a closer look at these five attributes.

2.2.1 Level and Behaviour of Autonomy

Autonomy is defined as the robot's ability to accommodate variations in its environment [Thrun (2004)] by sensing, planning, and acting to reach a goal without external control [Beer et al. (2014)]. Autonomy can range from manual teleoperation to full autonomy. Beer et al. (2014) have proposed a 10-point taxonomy to identify the level of robot autonomy along this spectrum. This is directly related to when and how the human or robot takes the initiative in HRI. Seizing the initiative can happen in a reactive, deliberative, or hybrid manner [Jiang and Arkin (2015)].

In a study related to this attribute, a hypothesis may choose the robot's level of autonomy as the independent variable and explore its effect on a range of dependent variables such as participant workload and robot perceived trustworthiness [Rau et al. (2013)]. The measurement of these dependent variables is further discussed in Section 5.1.

2.2.2 Nature of Information Exchange

The nature of information exchange is primarily defined by the communication channel and communication format [Goodrich and Schultz (2007)]. These can be affected by the distance between the human and robot in HRI, – for example, whether they are physically attached, in the same area, or in different areas [Agah (2000)]. Communication channels used for HRI often mimic those in human-human interaction. Such channels, and example formats, include visual, such as gaze [Admoni and Scassellati (2017)] or body gestures [Liu and Wang (2018)]; auditory, such as speech [Yongda et al. (2018)] or audio-based expressions [Frederiksen and Stoey (2019)]; and haptic, such as shared load [Chan et al. (2013)] or vibrations [Scheggi et al. (2014)]. A common strategy for enabling intuitive and effective HRI is to program robots to use communication channels and formats familiar to humans [Haddadi et al. (2013), Gielniak and Thomaz (2012), Parker and Croft (2010)].

Considering the first example in this section [Breazeal et al. (2005)], the main attribute of concern is the nature of information exchange. The hypothesis considers the communication method as the independent variable, comparing the combined use of implicit non-verbal behaviours and explicit communication, or the latter only. The dependent variables are robot understandability, task efficiency, and robustness to errors.

2.2.3 Structure of the Human-Robot Team

The structure of the human-robot team concerns the number (and type) of humans and robots, their roles, and the communication links among them. Directly relating to these factors, Yanco and Drury (2004) proposed a taxonomy that includes the ratio of people to robots, composition of robot teams, level of shared interaction among teams, and human role.

The human may take a range of roles: supervisor, operator, mechanic, bystander, or teammate [Scholtz (2003)]. The role of the robot can be identified by comparing HRI to human interaction with other agents, including other humans, animals, and objects and defining the role of the robot based on the similarity of its role to those agents [Collins (2019)]. This can also be affected by the robot's morphology [Yanco and Drury (2004)]. An anthropomorphic, zoomorphic, or functional robot is more likely to be deployed in roles matching its shape. In social robots, the ability of the robot to deceive has been suggested as an additional role capacity categorised based on the deception object, goal, and method [Shim and Arkin (2013)].

The structure of the team and the role of team members are among the independent variables that can potentially shape the basis of HRI hypotheses. Measurable dependent variables, for example, likability, trustworthiness, and team performance, are chosen based on a study's main research question(s). For example, Rosenthal-von der Pütten et al. (2013) examined humans' emotional reaction to the positive/negative treatment a robot receives in videos and based their hypotheses on existing psychological research. The independent variables were either manipulated within the experiment (e.g., different videos) or based on the user's personality traits. The outcomes (e.g., participants' emotional state and physiological arousal) were then evaluated via questionnaires or physiological measures.

2.2.4 Adaptation, Learning, and Training of People and Robots

Adaptation, learning, and training can happen for both humans and robots in HRI. Most HRI designs aim to develop intuitive interfaces for the human user. Therefore, robotic systems for HRI are often designed to minimise the amount of user training required [Goodrich and Schultz (2007)]. However, there is always some human learning and adaptation taking place [Green and Eklundh (2003)]. Sefidgar (2012) used robotic pets as therapeutic animals for treating anxiety by teaching patients to adapt their breathing to the robot's. Short-term HRI studies commonly treat learning in humans as an undesirable carryover effect to be mitigated. In longitudinal studies, adaptation and learning warrant careful study and may be the foci of the studies [Lee Koay et al. (2007)]. Certain high-risk application domains, such as bomb disposal, can benefit from or may require careful training of the user or operator [Rehfeld (2006)].

Robot adaptation has been used to personalise interaction and promote longer-term engagement with users [Ahmad et al. (2017), Tozadore et al. (2018)]. Robots can also learn tasks and skills from humans through various approaches, such as Learning from Demonstration [Argall et al. (2009)]. In the context of education, robots have taken on the role of teachers or peer tutors [Edwards et al. (2018)].

Training period and mode are examples of independent variables in a hypothesis that might affect various outcomes, such as user engagement and receptiveness. For example, a study on personal space examined people's behaviour (adaptation) around robots depending on past experiences, among other factors [Takayama and Pantofaru (2009)]. Their four hypotheses presented clearly how these independent variables are expected to influence measurable outcomes (e.g., personal space around robots).

2.2.5 Nature of the Task

The introduction of new technologies may change how people fundamentally perform a task [Goodrich and Schultz (2007)]. Hence, it is important to consider how people interact with a robot or perform a task when a new robot technology is introduced. Task type and task criticality (measuring importance of task outcome) [Yanco and Drury (2004)], as well as the task's cognitive and physical requirements [Rahimi and Karwowski (1990)], are introduced as part of the HRI taxonomy. Collins (2019) suggested that a robot be selected based on its suitability for a task rather than on the expected outcome. The level of autonomy of the robot is an influential factor in determining the type of tasks it can perform [Thrun (2004)], the aspects of the tasks it performs, and the extent to which it can perform those tasks.

The introduction of new methods and technologies, such as kinesthetic teaching [Wrede et al. (2013)] and augmented reality [Chan et al. (2020)], has fundamentally changed how humans can carry out tasks or interact with robots. For example, in a study examining methods for a robot programming task, the hypotheses related the independent variable of task shape (e.g., teleoperation or kinesthetic) to the outcomes of interaction ease and skill performance [Akgun and Subramanian (2011)].

2.3 Developing Research Questions with Stakeholders in Mind

As HRI is defined as "a field of study dedicated to understanding, designing, and evaluating robotic systems for use by or with humans" [Goodrich and Schultz (2007), page 204], it is crucial to maintain a human-centred perspective that centres HRI research questions around how people understand and interact with robots. Hypotheses to these research questions can then be constructed to be retested and replicated across studies. This includes in-depth and participatory analysis of the needs and expectations of both primary users and stakeholders of a robot; incorporation of the social, cultural, personal, and other various contexts of the user cases; and choice of suitable experimental design that measures metrics that reflect these needs and contexts.

For example, in an HRI study aimed at understanding how an assistive robot can better support older adults to promote ageing in place and quality of life, engineering or technology-focused approaches might lead to hypotheses that investigate the number or type of tasks a robot can successfully perform for a user but neglect the potential associated decrease in mental and physical health of the user [Oviatt (2021)]. Langer and Rodin (1976) found in a longitudinal study at nursing homes that residents who cared for a house plant themselves showed better overall health and a 10% lower mortality rate than residents whose plant was taken care of by the nursing staff. Adopting human-centred approaches, such as participatory design [Ostrowski et al. (2021)], can help facilitate mutual learning between the researchers and the stakeholders, which can allow researchers to formulate more suitable research questions and hypotheses that are better grounded in the context and support the development of more beneficial and ethical HRI systems.

It is important to consider the ethical implications of an HRI study from the start of its life cycle, by conceptualising the research question(s) and problem statement(s) concerning the potential benefits and risks an HRI system may have to its users. Moving away from the belief that the technology designer has the ultimate power in determining its usage, HRI ideation based on the particular circumstances of the users results in technology that is adaptable to diverse, real-life contexts [Søraa et al. (2022)]. Hypotheses should be beneficial not just to the researcher but to the larger community that might interact with the robot. This can be achieved through the formulation of hypotheses supported by evidence rather than focusing on the potentially biased assumptions of the researchers. For example, an interview with older adults about their interpretation of assistive robots intended for their age group highlighted the dominance of potentially negative stereotypes around age-related disability that shaped the robot design and discourse [Lee et al. (2016)].

To formulate hypotheses, stakeholders that are invested or interested in the work, or otherwise impacted by it, should be considered [Mathewson and Pilarski (2022)]. Similar to the approach already adopted in other fields, such as human-centred AI [Shneiderman (2022)], participatory design (see Section 11.1.7) – through communication with stakeholders, gathering their ideas, hearing their concerns, and understanding how they would evaluate the system – can be adopted to develop ethical hypotheses in line with the actual needs of those involved. With the advent of robots across the globe, ethical formulation of hypotheses goes beyond the functions of technology and users' age group considerations. It now covers cultural [Lim et al. (2021), Bröhl et al. (2019)], and socioeconomic background [Singh et al. (2022)], and religious beliefs [Weng et al. (2019)] of the stakeholders, among other factors. Thus, interdisciplinary collaboration and consultation with stakeholders can benefit the development of ethical and inclusive hypotheses in HRI research. Thus, HRI researchers need to recognise this high level of interdisciplinarity through increasing disciplinary inclusivity in HRI research and reproducing work from outside disciplines with the involvement of the original researchers [Gunes et al. (2022)].

An example of formulating hypotheses that address diverse stakeholders and ethical challenges in HRI is robots designed for aged care and healthcare facilities (e.g., Riba, the patient-lifting robot) [Sharkey and Sharkey (2012)]. On one hand, a hypothesis centred around how to use robots to reduce the workload of healthcare providers or caretakers (i.e., providers as the primary end-users) may lead to patients feeling objectified, where patients are treated not as persons but as heavy objects. Hence, additional considerations are needed to respect the dignity and independence of the patient. On the other hand, a hypothesis centred around patients without considering healthcare providers may elicit negative responses, such as human workers feeling threatened that the robots will be an extra burden or will replace their jobs. In addition to primary and secondary users, the broader societal context can also influence the potential benefits or risks of HRI systems. For instance, when introducing home assistant robots for lifting and carrying tasks for elderly people, one needs to take into account that the human carers originally performing these tasks often provided benefits beyond the task itself, including social interactions that can reduce feelings of isolation [Sharkey and Sharkey (2012)]. This example highlights how ethical values, which will be discussed in Section 8.3, impact the conceptualisation of HRI research.

2.4 Pre-Registration of Hypotheses

Pre-registration of research questions, hypotheses, and study design can promote ethical HRI research by improving transparency. It applies to both confirmatory and exploratory studies. In some fields, such as psychology, pre-registration of hypotheses or the research plan before conducting the study is recommended to avoid bias, as well as to ensure that the methodology of the study is justified based on its hypotheses, rather than the study outcome [Toth et al. (2021)].

Pre-registration can provide benefits such as ensuring both positive and null results are published and increasing replicability and reproducibility. As such, it also helps to ensure the ethical practice of research and that less favourable research results are also published and not selectively hidden. Although pre-registration does not guarantee true reports, it reduces false positives, hidden confounds, and fraud by improving transparency [Simmons et al. (2021), Toth et al. (2021)]. A good way to improve transparency, replicability, and reproducibility is also to reduce the emphasis on significance testing (Section 6.3.1) and to enforce or encourage the sharing of data, analysis code, and study materials. For example, for their study, Laban et al. (2021) made de-identified datasets, stimuli, and analysis codes from their study freely available online, to enable and encourage other researchers to test alternative hypotheses and more exploratory analyses. With online cloud storage platforms and code repositories becoming increasingly available and lower cost, the sharing of code and datasets has also become more common.

Pre-registered hypotheses can guarantee subsequent publication of the study regardless of results in certain fields. While pre-registration is not yet common practice in the field of HRI, it may be beneficial for the field to consider moving towards that direction, especially for confirmatory studies. Shifting the focus of reviews from minor imperfections to the empirical and conceptual evidence presented can also support major innovation. However, once published, a pre-registration may restrict the researcher from making any subsequent changes to the study design, even if such changes may be beneficial (for example, selecting a better analysis method or increasing the sample size), as they are then expected to carry out their study exactly as described in the pre-registration. Since pre-registration can reduce flexibility in the face of unforeseen challenges and exploration of data, it may be less suitable for exploratory studies and more suitable for confirmatory studies [Toth et al. (2021)]. Furthermore, pre-registration needs to be adjusted following initial pilot work and thus cannot always be completely pre-specified before a viable research protocol is finalised.

Researchers can register their hypotheses at websites such as the Center for Open Science (`https://osf.io/prereg`), AsPredicted (`https://aspredicted.org`), or the U.S. National Library of Medicine (`https://clinicaltrials.gov`). Results are then reported with reference to the initial hypotheses to avoid a selective report based on the study outcome. Some journals (e.g., *JMIR Research Protocols*) may automatically generate an identifier for submitted and published research protocols or proposals. The identifier is then linked to all subsequent result papers. While registering hypotheses is not yet common practice in the field of HRI, it may eventually be adopted for the benefits mentioned in this section.

2.5 Summary

The two main types of HRI studies, exploratory and confirmatory, were introduced. We focused on confirmatory studies in this section, in which identifying the research question and corresponding hypothesis marks the beginning of the experimental design process. A good hypothesis is based on a sound background and testable with an experiment. It needs to be designed to shed light on the research question. A clear hypothesis concisely relates the independent variables to the dependent variables. We suggest the five discussed HRI attributes as the starting point for identifying the independent variables. These variables are manipulated in the experiment to study their effects on the outcome. Once the independent variables have been identified, the set of values to be tested and thus the dependent variables can also be decided.

Formulating a hypothesis is a critical initial step in any confirmatory HRI study. The hypothesis sets the ground for the next steps of the study, from experimental protocol and design of evaluation methods to recruitment of participants, evaluation metrics, statistical analysis, and reporting the outcomes.

Stakeholder engagement and the generation of research questions and hypotheses based on their needs and the HRI application context are required for the ethical design of HRI studies. The impact of an HRI system on the stakeholders within the use context can be validated through an iterative cycle of exploratory and confirmatory studies. Furthermore, hypotheses pre-registration can benefit the transparency of HRI research. However, it may also reduce flexibility in addressing unforeseen challenges.

Recommendations for Forming HRI Research Questions and Study Hypotheses

- Confirmatory HRI studies are useful for testing hypotheses and uncovering the relationship between isolated factors and the interaction outcomes or behaviours. However, they can be limited in understanding the overall dynamic and complexity of HRI. Exploratory HRI studies on the other hand are helpful for generating new research questions and hypotheses embedded in the interaction context, while observations made from these studies require further analysis and replication to ensure their validity and robustness.
- When constructing HRI study hypotheses, it is useful to identify the various design factors that can be altered to affect HRI and how they impact the user and the interaction, possibly through a hypotheses-generating or exploratory approach. However, for a confirmatory study, it is also important to constrain the scope to focus on the factor(s) of interest and to ensure the feasibility of the study.
- Pre-registration of hypotheses can help improve transparency and avoid bias, which is essential to ethical HRI research. However, it may reduce flexibility when encountering unforeseen challenges. Hence, this is more suitable for confirmatory studies. Publications, reviewers, and editors can help encourage pre-registration of HRI studies.
- Forming a good hypothesis:
  - A hypothesis is developed to formulate a testable claim to answer the research question.
  - A hypothesis is a predictive statement on how an experimental condition(s) is expected to affect a measured outcome(s).
  - A good hypothesis is testable and concise and names key constructs.
  - When constructing experiment hypotheses for HRI studies, it is useful to consider the various design factors that can be altered to affect HRI.
  - To improve rigour, the experimenter may consider registering the hypotheses before conducting the study.

3

Study Context and Experimental Design

With research questions and hypotheses formulated, the next step is to determine the experimental conditions, the components of the study, and the evaluation methods that are appropriate to explore them. HRI experimental conditions may include a series of tasks, sessions, or interaction patterns depending on the hypothesis being investigated. This chapter explains the different study components (e.g., study design structure, choice of location, choice of robots) in relation to HRI.

HRI studies are often built to simulate or replicate an intended use context with the goal of understanding how robots may be deployed in real-life scenarios. For example, the use of service robots in malls [Kanda et al. (2009)] or therapeutic robots in long-term care facilities [Moyle et al. (2017)]. Considerations should be given when building the study context to ensure that suitable approaches are used. This includes designing the robot's appearance and behaviours to avoid reinforcing negative stereotypes, recruiting participants that are diverse and representative of the intended users, and employing suitable screening and cleaning steps in data collection. Here, we discuss considerations for study design and context, including planning for failures to reduce potential risks in HRI studies.

3.1 Study Design

Study design determines how the participants are assigned to the different experimental conditions. Common designs include between-subjects, within-subjects, and mixed-model factorial designs (combining both between- and within-subjects designs, not to be confused with the mixed-method evaluation and data analysis, that is, combining quantitative and qualitative evaluation methods, which will be introduced in Chapter 12).

Within-subjects design is used to understand how the same group of people perceive different factors in HRI (e.g., a robot's communicative

DOI: 10.1201/9781032693682-4

behaviours) or how repeated exposure to a factor changes their perception and interaction with a robot over time. The design between subjects examines how HRI outcomes are impacted by a specific factor that differs between each group of participants. Mixed-model factorial designs include both within- and between-subjects factors. They are used to compare one set of conditions with another set, while also comparing the individual conditions within each of these sets [Christensen (2007)]. Different study design methodologies may be suitable depending on the study objective or the availability of participants. In addition, there is often the consideration of including a control group or a baseline condition in the study design. In this section, we will explain these different study designs in more detail.

3.1.1 Within-Subjects Design

In a within-subjects design, there is one group of participants. Each participant is exposed to all the experimental conditions. This experimental setup minimises experimental noises from individual differences and can be useful when large differences are expected (e.g., in speech or language behaviours). A smaller sample size can be used in a within-subjects design, which is helpful in situations where participants are difficult to recruit [Lazar et al. (2017)]. Other advantages include the possibility of performing experiments in a single session sequentially (i.e., more time efficient) and smaller statistical noise (e.g., the scores of the same participant are compared between each experimental condition) [Christensen (2007)].

The challenges with a within-subjects design stem from the long duration of the experiment and repeated exposure to the task. A longer experiment time can increase the likelihood that participants are affected by non-intentional incidents. HRI-related examples include program errors or glitches [Krauth (2000)], or longitudinal changes that can influence robot algorithms, such as lighting in the room or environmental noise. This can create a greater diversity of individual experience between conditions. Repeated exposure to tasks can also lead to participants anticipating the outcome of the experiment, developing habituation, improvements through practice, and/or fatigue, which can all influence task performance. In a study that examined habituation and robot autonomy, the authors found that participants who have adapted to robots would increase their control over the robot (i.e., decrease robot autonomy) to avoid unexpected actions [Koay et al. (2007)].

Repeated exposure to tasks can also change the psychological aspect of HRI, such as through the mere exposure effect [de Graaf et al. (2016), Paetzel et al. (2020)]. For instance, the preceding studies found that the participant's perception towards the robot improved not due to changes in the robot's functionality, but rather due to participants becoming more familiar with the robot. Within-subjects design is further prone to cross-condition contamination, where exposure to one condition context affects the participant's response during the second condition, sometimes known as the carryover effect [e.g., Yamashita et al. (2016)]. Ordering effects can also occur based on the sequence of conditions presented [Salkind (2010)]. Researchers can consider using a Latin square design and/or randomisation to control the impact of habituation and other confounding variables and incorporate the necessary breaks between experiments [Lazar et al. (2017)]. Other considerations such as trial location and session duration are further discussed in Section 3.2.

In summary, within-subjects designs are best used when researchers intend to evaluate a measure across multiple time points or test the effects of different conditions. It is also suited when the sample size is limited. However, using the within-subjects design requires careful consideration on the possibility of sequence effects, fatigue, and practice effects.

3.1.2 Between-Subjects Design

In between-subjects designs, there are two or more groups and each participant is allocated to one group only. Participants in each group are then exposed to one experimental condition. This could be one task, session, or interaction pattern. The number of participants in each group is determined based on the number of conditions. Compared to within-subjects design, a strength of between-subjects design is that it minimises confounding variables such as learning effect, fatigue, and frustration that occur as by-products of running long experimental sessions [Christensen (2007), Lazar et al. (2017)].

While between-subjects design can avoid some of the negative effects of the within-subjects design, since participants are different between each experimental condition, the collected data will be affected by individual differences without randomisation (i.e., random allocation of participant groups to experimental conditions). As a result, it can be more difficult to detect significant differences, and type II errors (i.e., false negatives) are more likely to occur [Johnston and Pennypacker (2010), Bethel et al. (2020b)].

In a between-subjects design, the experimental conditions should be distinct from each other for between-group differences to be perceived, otherwise the researcher runs the risk of detecting no differences. Furthermore, a larger sample size is required to mitigate the impact of individual differences. Additional discussion on the trade-offs of different study designs can be found in Lazar et al. (2017), Krauth (2000), and Christensen (2007).

In general, between-group design is best used for testing the difference between conditions, behaviours, or tasks, without having a participant exposed to similar conditions multiple times. This design can strongly identify across-groups effects compared to within-group effects. Nevertheless, between-subjects design often requires a larger sample size to produce statistically significant results, and appropriate participant randomisation is required to reduce the influence of individual variances.

3.1.3 Mixed-Model Factorial Design

A mixed-model factorial design utilises both between-subjects and within-subjects designs, where the between-subjects aspect is used to investigate multiple independent variables while other random effects are explored using within-subjects designs. This study design can simultaneously explore the effect from individual variables and the interaction effects [Lazar et al. (2017)]. The limits of each study design structure discussed previously still apply, and additionally, mixed-model designs often require more participants to properly randomise the study. Mixed-model designs are best suited when researchers are seeking to assess both between-group and within-group effects. Mixed-model designs require a large sample size and a thorough experimental protocol design to implement correctly but produce rigorous results.

3.1.4 Baseline and Control

A control is used to demonstrate how another group or condition performs against a neutral or baseline condition. This can be important for experiments that aim to evaluate the impact of a new technology from different perspectives, such as usability, functionality, and level of engagement. An example includes an immediate session compared to a delayed session. Commonly used control conditions include wait-list/delayed (e.g., participants receiving no interventions delivered by robot) and active control (e.g., participants interact with a monitor instead of a robot or interact with a robot following a fixed scripted behaviour instead of an adaptive behaviour). Control scenarios can take many forms, such as comparing the effect of a robot-delivered treatment with the control scenario of a human delivering the same treatment, or an animal treatment, or conventional treatments. Interested readers can learn more about control conditions in experimental methodology books [Allen (2017), Price et al. (2015)].

In HRI studies, there are many options to benchmark robot performance. Control may not be necessary in an initial study, but later studies that seek to substantiate the claims in more detail require comparisons against a neutral or baseline condition.

3.2 Study Context

To ensure ecological validity of an HRI study, careful considerations are required in the study context. In this section, we discuss four aspects of the study context, namely, trial location, session frequency and duration, robot type, and task behaviour. In addition, as unintended failures can occur due to technical limitations of robots or study complexity, we discuss considerations on planning for failure and contingency in the experimental design.

3.2.1 Trial Location

A variety of trial locations can be considered with different suitability for a study design. In 2016, 75% of HRI studies were conducted in a controlled lab environment [Baxter et al. (2016)], while recently the field is moving towards more realistic settings and in-the-wild studies. Studies conducted in artificial environments might be susceptible to the Hawthorne effect or the white lab coat effect, with participants behaving differently while being observed. Additionally, HRI studies are often limited to a single session of short-term exposure, making the results susceptible to the novelty effect [Reimann et al. (2023)].

Laboratory testing involves conducting tests in a controlled space, such as a research laboratory or office space of the research team. Existing HRI studies are most frequently conducted in a laboratory setting due to the ease of testing, convenient access to robotic equipment, and rigorous control and observation capabilities. It is generally easier to control for unforeseen variables in laboratory testing, but observed interactions might be less natural and lack real-world relevance. In addition, only participants with access to the laboratory are able to attend the session (i.e., those who live nearby or have immediate access to a university campus). Fundamentally, physical laboratory testing could lead to loss of fairness, as the representation may be skewed toward the local, convenient sample.

Field testing involves the deployment of robots outside controlled laboratory environments, in either public settings such as museums [Pitsch et al. (2009)], shopping malls [Kanda et al. (2009)], and schools [Lemaignan et al. (2016)], or in non-public environments such as industrial spaces [Puljiz et al. (2018)]. The HRI community is encouraging field tests in realistic settings that better represent the final deployment use case of the system being evaluated, that is, in-the-wild HRI studies. However, this can be difficult due to access or resource constraints and may be at the expense of rigorous and structured testing methods controlling for confounding variables.

Online environment testing involves conducting the experiment through a digital medium, such as using telepresence, online crowd-sourcing platforms, or computer-based mediums and simulations. Online environments facilitate an expanded sample size of participants. Such remote HRI study methodologies are increasingly adopted, and were especially accelerated during the COVID pandemic [Gittens (2021)]. Online HRI studies were shown to elicit behaviours, such as upper body communicative gestures [Honig and Oron-Gilad (2020)], and perceptions, such as intention of use [Babel et al. (2021)], in remote participants, comparable to HRI studies in person. However, without direct interactions, online studies may be better suited for measuring perceived metrics or expectation, such as perceived trust of a robot's behaviour. In video studies, it is also important to provide clear instructions, for example, if the participant is expected to respond to their own interpretation of the video or respond on behalf of the person or actor in the video. The conduct of studies online using crowd-sourcing tools is described in more detail in Section 4.3.

With recent advances in simulator environments and cloud computing, there have been online HRI experiments using interactive simulators, such as in social navigation [Tsoi et al. (2021)] or simulated warehouse scenarios [Esterwood and Robert Jr (2023)], that have increased the incorporation of contextual and interactive factors in remote studies. Moreover, the development of robot functions may happen in a simulator or on benchmarking datasets without conducting a user study to evaluate them in direct interaction with participants [Savva et al. (2019), Wiederhold et al. (2024)]. However, accurate modelling of human behaviours, physical aspects, and diverse environments remains challenging in simulation-based HRI research, known as the sim-to-real gap [Höfer et al. (2021)].

In summary, HRI studies can take place in the laboratory, in the field, or online. There is always a trade-off between a controlled laboratory environment and the real world, and there will be differences between online versus physical in-person interactions. The selection of the appropriate trial location depends on the hypothesis under investigation. Researchers can also work incrementally, by first conducting studies in the laboratory and then moving towards a more realistic setting, with additional large-scale validation online. Trials conducted in one setting should not be over-generalised in the abstract, introduction, or discussion sections during reporting as contributing new knowledge across all settings.

3.2.2 Session Frequency and Duration

To determine the duration and frequency of the test session, it is important to consider the following parameters:

- How many conditions?
- How many sessions?
- How long is each session?
- How long is the duration of the whole experiment?
- How often do sessions run?
- What time is best for follow-up studies?

The parameters related to session duration and frequency can be determined given the hypothesis and independent variables of the experiment. In addition, it is desirable for researchers to conduct a pilot proof-of-concept trial with a limited number of participants to help determine the frequency and duration of the session before larger-scale recruitment. An increase in the number of sessions and the duration of the study can help reduce the impact of the novelty effect (i.e., positive perception of a new technology spikes on first encounter) and the study duration needs to be designed such that the effect of the experimental condition can be measured after the novelty effect attenuates. Example studies that need to avoid the novelty effect can be found in Sung et al. (2009) and de Graaf et al. (2016). However, trials that run for a long time can present challenges in participant retention [Fernaeus et al. (2010)].

3.2.3 Type of Robot

The choice of robot depends on how the different capabilities of a robot align with the requirements of an experiment. Examples include selecting a robot based on its appearance, mobility, processing power, and level of functionality. There are a limited number of robotic systems that are commercially available, and not all the systems have the features that one might be interested to explore (e.g., facial features). The selection of robots should take into account how the target user group will perceive the appropriateness of the robot and the intended use case. Alternatively, Patten and Newhart (2017) provided a list of design principles for building a customised robot.

The appearance of the robot can affect how people perceive the robot's sense of intelligence, sociability, likability, credibility, and submissiveness, along with other attributes [Phillips et al. (2018)]. A humanoid robot can be intuitive to interact with, but the physical similarity to humans might set a higher expectation for the robot's capabilities [Patten and Newhart (2017)]. Zoomorphic robots that look like animals can be less intuitive but free from a user's pre-conceived expectations of human-like behaviours and capabilities. Trade-offs are often present between a more anthropomorphic robot compared to a non-biomimetic one. For example, social robots used in autism spectrum disorder therapy can take on various forms, such as robots with human-like facial features, stylised miniature human dolls, zoomorphic robots, mechanical robots, and robots with cartoon-like features [Scassellati et al. (2012)]. Robots that are not restricted by human-like characteristics can help exaggerate social signals, which help autistic children focus on learning the desired skills without the presence of confounding cues [Scassellati et al. (2012)].

3.2.4 Task Behaviour of the Robot

A robot's task behaviour is often the centre of the experimental design, including defining how the robot will behave and what tasks it will perform in the experiment. Example tasks may be picking up an object for a person, passing a group of people in a narrow hallway, or communicating a set of instructions via conversation. The type of task behaviour is dependent on the expected outcomes and the variable of interest investigated in the research questions. The robot can perform these tasks using different autonomy levels, as was described in Section 2.2. The robot's task behaviour can also change or evolve during the experiment through the operator's active tuning (i.e., adaptable robot) or the robot's active learning (i.e., adaptive robot) to better tailor the robot's assistance to the operator's need [Schneider and Kummert (2020)].

In teleoperation tasks, a robot's task behaviour might be fully dependent on the human operator. When the robot's role in the experiment has some autonomy, the robot behaviour can be implemented algorithmically or partially controlled by a human operator. If the participant in the experiment is unaware that the robot is remotely operated, this is known as a *Wizard-of-Oz* (WoZ) experiment. This means that the participant is led to believe that the robot is fully capable of making its own choices and actions within the experiment, while in reality some or all of the decisions are being made by a hidden human operator.

WoZ is a useful approach, especially for circumventing the technical constraints of robots and answering research questions about a proposed robot system that is not yet available or possible. For instance, WoZ can be used to let an operator identify a pedestrian who shows interest in interacting with a robot and investigate the effects of different non-verbal communicative behaviours of the robot for engaging the person. However, special considerations must be taken when using WoZ as it does not reflect how a robot can behave fully autonomously in future deployment. Furthermore, the WoZ approach involves deceiving participants. To avoid participant confusion or frustration, suitable debriefing needs to be conducted after the experiment to ensure the participants fully understand which robot behaviours were autonomous versus teleoperated and why WoZ was adopted. When possible and feasible, it is best to employ robot behaviours that the robot can perform independently to avoid deception or assessment of effects that are not yet achievable. On the other hand, WoZ is most suited when evaluating an HRI factor that cannot be delivered on its own for the purpose of theoretical exploration or when the robot is used as an intentional avatar for experiments. Wizard training and other ethical considerations regarding the WoZ technique both from the operator's and participant's perspectives are further explored in Section 4.5.2.

3.2.5 Planning for Failure and Contingencies

It is critical to account for the possibility of failure and contingencies during experimental design. Common failures include equipment failure and deviations from the experimental protocol, which can be mitigated, to some extent, through pilot studies (see Section 6.1). Other approaches include a contingency protocol on what to do in the event of a failure or anomaly with the robot. Besides technical failures, robots can also violate social norms by accident, leading to changes in the user's perception of the robot outside of experimental manipulation, that is, "social errors" [Tian and Oviatt (2021)]. Recovery strategies from different failures can include improving the explainability of the robot's action [Tabrez and Hayes (2019)] or adapting the robot's behaviour in response [Tian and Oviatt (2021)]. Recently, the concept of imperfect HRI has been discussed in this context, which highlights the dynamic nature of HRI research. Finally, comprehensive data management plans help to avoid data loss challenges, including backup equipment and data capture whenever possible. Further steps to mitigate failures in a trial evaluation through data cleaning will be discussed in Section 6.2.1. Beyond equipment and data, failures can arise from human factors and biases, which we will further discuss in Section 10.1.

Planning for failure is an important step in experimental design due to the dynamic nature of experiments involving human participants and the various hardware and software of robots. Researchers are advised to conduct pilots and test runs prior to the experiment to identify and minimise the risk of potential failures. Participants of these pilots may either be a small sub-sample of participants that are discarded from the final analysis (e.g., the first three trials) or be members of the research team who complete the session without their data formally collected, as will be further discussed in Section 6.1. In addition, researchers may conduct feasibility studies with target users as part of the iterative design process [e.g., Jordan et al. (2013)], which will be further discussed in Section 7.2.

3.3 Summary

We described common study designs in an HRI study and their corresponding advantages and limits. We also discussed how contextual factors can be incorporated into the experimental design of HRI. It is desirable for an HRI study protocol, such as the location or duration of the interaction, to resemble the intended interaction scenario as closely as possible. However, when this is difficult to achieve due to feasibility or risks in the experiment, differences in the experimental context should be clearly reported and considered when interpreting findings of the HRI study.

Recommendations for Study Design

1. **Within-subjects designs** are best used for evaluating a measure across multiple time points to test the effects of different conditions while minimising noise from individual differences. It is suitable when a limited sample of participants is available. Note the possibility of sequence effects, fatigue, and practice effects.

2. **Between-subjects designs** are best used for testing the difference between multiple conditions, behaviours, or tasks, while avoiding the participants being influenced by similar conditions. It is suitable when the expected impacts of individual differences on the research questions are small. Employ appropriate randomisation techniques to randomly assign participants to each group. Using a large sample size is required to produce statistically significant results.

3. **Mixed-model factorial designs** are best for assessing both between- and within-subjects effects. Mixed-model designs require a larger sample size and careful design of conditions and study protocols to implement correctly.

4. **Control or baseline conditions** help to provide a comparison for the research conditions of interest. In HRI studies, there are many options for implementing a baseline. A control condition may not be necessary in an initial or exploratory study, while confirmatory studies and studies that are looking to substantiate claims in more detail can benefit from using one.

Considerations of HRI Study Context

1. HRI studies can take place at various **locations**. There is always a trade-off between a controlled laboratory environment and the uncertain real world, and there will be differences between online versus in-person interactions. The appropriate trial location depends on the hypothesis under investigation. Researchers can also work incrementally, by first conducting studies in the laboratory and then moving towards a more realistic setting. Trials conducted in one setting do not automatically generalise across all settings.
2. HRI experiments can include interactions happening at different **session frequencies and durations**. The novelty effect can influence outcomes of experiments conducted within a short time frame, while long-term studies can pose difficulties for participant retention. The appropriate session frequency and duration depends on the hypothesis and may be iteratively adjusted to better align with the intended application scenario.
3. When possible and feasible, it is recommended that researchers use the **type** of robot that resembles the intended use case and test **task behaviours** that are achievable during deployment. The Wizard-of-Oz approach may be used when evaluating a task behaviour that cannot be delivered on its own for the purpose of theoretical exploration.
4. **Failures** can occur during experiment and data analysis due to human or technological factors. To identify and mitigate potential failures, it is recommended that researchers run multiple pilots prior to participant recruitment and performing the experiment.

4

Identifying and Recruiting Participants

HRI research envisions robots interacting or collaborating with users in a variety of settings. A major requirement to validate proposed approaches is the participation of users during experiments. Therefore, recruiting the appropriate participant population is crucial. In this chapter, we first identify the types of participants involved in HRI studies and summarise the important steps in participant selection. We then discuss the procedure for participant recruitment, and retention, and highlight various ethical considerations pertaining to the people involved in HRI studies.

4.1 Roles and Types of Participants

People involved in an HRI study may be facilitators of the study, such as the researchers themselves, Wizard-of-Oz (WoZ) operators, and confederates. They may also be participants who represent users and stakeholders, such as interactors, observers, teleoperators, and bystanders. WoZ operators are those who operate the robot remotely as if it were a fully functional system [Riek (2012)]. Confederates are actors or researchers who pretend to be experiment participants but in reality work for the research team [Allen (2017)]. The use of WoZ operators or confederates allows researchers to simulate scenarios and measure effects that may only rarely arise in naturalistic interaction. Interactors usually represent the primary users of an HRI system and are often at the centre of designing and evaluating an HRI system, especially in human-centred approaches. Observers are those viewing the interaction sessions either in person or as videos and interpreting the content based on HRI evaluation metrics. Teleoperators are those controlling the robot directly as part of a human-robot collaboration task and are also considered the primary users of the system. Finally, bystanders are usually secondary users, that is, people whom HRI systems may indirectly impact, such as pedestrians sharing the same road with autonomous vehicles [Saleh

DOI: 10.1201/9781032693682-5

et al. (2017)] or occupants or passers-by of a public space shared with a robot [Rosenthal-von der Pütten et al. (2020)]. While bystanders may not be the interactors or primary users of an HRI system, their behaviours may influence the perceptions and actions of the primary users [Tan et al. (2018)] or the social acceptance of the HRI system [Scholtz and Bahrami (2003), Salvini et al. (2010)], making them stakeholders of HRI studies.

HRI studies more often account for the stakeholders who are positively affected, for example, input from healthcare providers on a therapeutic robot. However, parties who can negatively impact the interaction (e.g., cybersecurity hackers) may be included to provide new perspectives [IEEE (2021)]. As the ethical values of different stakeholders can conflict with one another, it may be necessary to prioritise them, as discussed in Section 8.3. Further, a suitable experimental design is required to validate that selected ethical values are met (see Section 8.4).

Identifying stakeholders, that is, people who may be affected by HRI systems [Bryson (2004)], is the prerequisite to understanding and validating their ethical values and requirements. The majority of current HRI studies focus on users who directly or indirectly interact with robots as the main stakeholders. Engaging other stakeholders, such as policymakers or the managerial team of a deployment site, is important, especially for the commercial application of HRI. However, here we focus on HRI study participants who represent primary or secondary users of HRI systems and who are the most impacted by the research outcomes.

4.2 Participant Selection

4.2.1 Sampling

Sampling is the process of identifying a group of participants for conducting an HRI study. A representative participant sampling facilitates ethical considerations, including fairness, equality, and beneficence. While unbiased samples through random sampling are preferred, biased (i.e., non-random) sampling techniques, for example, convenience sampling and purposive sampling [Etikan et al. (2016)], are often used due to practical constraints, such as participant accessibility, proximity, and availability [Etikan et al. (2016)]. The majority of existing HRI studies use convenience samples involving university students [Bartneck et al. (2020b), Baxter et al. (2016)]. The use of convenience samples generally reduces the external validity of the research findings as the participant population is not representative of the user population in terms of social traits, cognitive performance, and attitudes towards technology [Bartneck et al. (2020b), Baxter et al. (2016)]. Geographic proximity is often another reason for using a convenience sample, which comes with the trade-off of limiting the cultural diversity of the participants. In general, convenience samples introduce sampling bias (e.g., excluding certain groups) or sampling error (e.g., including a higher proportion of certain groups) [Patten and Newhart (2017)]. When using convenience sampling (e.g., university students), researchers might consider balancing potential confounding variables, such as gender or educational background, during participant selection [Bartneck et al. (2020b)].

In purposive sampling, participants are deliberately chosen for certain characteristics that they possess, such as knowledge or experience that would enable them to assist with the research. There are different types of purposive sampling methods, including maximum variation sampling, homogeneous sampling, typical case sampling, extreme/deviant case sampling, total population sampling, and expert sampling, as detailed in Etikan et al. (2016).

Sampling affects the statistical power of the study, which is determined based on the combination of the number of trials, the number of participants, and the number of random or uncontrolled factors. A lack of statistical power can lead to both type I errors (false positives) and/or type II errors (false negatives) and result in wrong conclusions being made. Sampling more participants from a diverse population is the standard and most straightforward way to increase statistical power [Dix (2020)]. More discussion on the sample size can be found in Section 10.3.

4.2.2 Randomisation

Randomisation and stratification can help assign participants to groups, conditions, and observers while minimising biases. Randomisation involves assigning participants with an equal chance to experimental groups to explore the effects of the condition and prevents researchers from predicting the results of the conditions. Simple randomisation is a basic form of condition assignment, which can involve methods such as a coin toss or simple computer-generated function [Beller et al. (2002a)]. Simple randomisation can be used for sample sizes above 200 participants, given their statistical likelihood to even out in terms of allocation, but cannot be used for less than 100 participants as there might be an unequal number of participants in each group [Kang et al. (2008b), Beller et al. (2002b), Kang et al. (2008a)].

Stratified randomisation, or "matching", can be used when the characteristics of the participants need to be equalised across the experimental groups. For instance, if age is a factor that can affect the HRI experiment, researchers can stratify the participants to ensure that similar age groups are present in each participant group. However, factors relevant to the experiment need to be identified and normalised beforehand, which might not always be possible. There are other stratification techniques such as block randomisation [Doig and Simpson (2005)] or response adaptive randomisation [Higgins (2003)], which is a technique to adjust the condition assignment favouring conditions with better performance based on ongoing data collection. While these techniques are more common in the medical field and less common in current HRI studies, they can be useful in longitudinal studies with socially assistive robots.

Stratification requires separating participants based on certain characteristics, which may bring issues of representation and marginalisation, for instance, using a binary representation of gender. In medicine, stratification based on race/ethnicity is often based on categories suggested by the US census [Lee (2007)]. However, the definition of the categories may inherently bias the more powerful and visible groups [Lee (2007), Berente et al. (2011)]. The first step towards addressing this ethical challenge is to ensure that the classification being used for stratification aligns with the research aims. For instance, in a previous study evaluating the PARO robot as a treatment for emotional and behavioral symptoms of dementia for people living in long-term care facilities [Moyle et al. (2017)], participants were first stratified based on the organisation type (private vs. nonprofit) and then randomly assigned to one of the experiment conditions. Stratification helps minimise the effect of confounds and ensures representative samples are selected from the stratified groups. Other forms of classification besides socioeconomic factors should also be considered, such as grouping using other shared attributes or shared goals [Berente et al. (2011)].

Selecting representative, diverse participants allows for different perspectives to be heard and considered. Further design approaches [Costanza-Chock (2020)] can be taken to intentionally elevate underrepresented or disadvantaged groups to improve the fairness and inclusion of HRI studies, such as inviting marginalised user groups as participants of participatory design sessions to prioritise their values and functional requirements for the HRI system.

4.2.3 Blinding and Allocation Concealment

Blinding and allocation concealment are techniques that are used to further remove bias effects in experimental condition assignment and are decided on during the study design stage [Wood et al. (2008), Ruxton (2017), Schulz and Grimes (2002)]. Blinding is the process of preventing the participants, researchers, or both parties from knowing what intervention or experimental content participants are allocated to. Conventional blinding strategies include single-blinded or double-blinded studies. Blinding is often not reported [Hróbjartsson et al. (2009)], and researchers are encouraged to include this information as part of the study design.

Allocation concealment is when the researcher does not know the upcoming experimental condition until the moment of assignment. Blinding reduces the effect of observer bias (i.e., the observer's judgement is influenced by the knowledge of the group), whereas allocation concealment reduces allocation bias (i.e., how participants are assigned to each group) [Ruxton (2017)]. Both blinding and concealment can be executed cheaply. For instance, researchers studying the effect of different robot appearances can ask a researcher unrelated to the project to randomly assign letters to each robot. The mapping between the robot and the coded letter is concealed from the researchers who are analysing the data until the data has been processed. Allocation concealment can be implemented using paper-based techniques [Doig and Simpson (2005)] or using online tools such as Study Randomizer [Phase Locked Software (2021)].

4.3 Participant Recruitment and Incentivisation

4.3.1 Recruitment Methods

During participant recruitment, the required sample size, the best recruitment venue, and ways of motivating participants to join the study are among the key considerations. The right sample size will depend on the study design structure and the desired statistical power.

Participants can be recruited through contacts (e.g., phone calls, email, or social media), community groups, mailing lists, volunteer banks, schools, external collaborators [Lazar et al. (2017)], and crowd-sourcing platforms, for example, Amazon Mechanical Turk [Chandler et al. (2014, 2015)] or Prolific [Palan and Schitter (2018)]. The techniques and the corresponding recommendations are summarised in Table 4.1.

TABLE 4.1
Participant Recruitment Methods

Method	Advantages	Disadvantages
Contacts	Participants can be recruited via the researcher's external collaborators, such as recruiting patients through collaborating therapists [Lemaignan et al. (2016)] or organisations [Tapus et al. (2009)]. Researcher contacts are more likely to agree to participate.	This technique might be susceptible to snowball sampling, which is a biased sampling technique where one participant refers to other potential participants [Patten and Newhart (2017)]. Participants who are contacts of the researchers might be less likely to provide negative feedback.

Continued on next page

TABLE 4.1
Participant Recruitment Methods (Continued)

Method	Advantages	Disadvantages
Community groups, professional organisations	This method makes it easier to contact a large group of specialised participants at once (good for purposive sampling).	Response rates might be low and lead to a biased sample. Researchers can consider running a short presentation at the corresponding organisation to raise awareness and gauge interest.
Mailing list	This recruitment method is cheap, fast, and easy to execute.	Response rates might be low and lead to a biased sample. Cold emailing might elicit a negative perception towards the study, and hence it is preferred to recruit through existing mailing lists.
Volunteer banks	Volunteer banks enable access to a large number of participants.	Volunteers might be fundamentally different from non-volunteers, leading to sample bias. For instance, volunteers might be more excited about technology, have prior experience with robots, or be more willing to adopt new technologies compared to non-volunteers [Patten and Newhart (2017)].

Continued on next page

TABLE 4.1
Participant Recruitment Methods (Continued)

Method	Advantages	Disadvantages
Crowd-sourcing platforms	This method is also cheap, fast, and easy to conduct, as the experimental tasks typically involve viewing pictures or videos online to replace in-person HRI [Dautenhahn (2018)]. Different filters on the platforms or sites also enable researchers to include a more diverse population or find participants with specific skills [Chandler et al. (2014)].	Crowd-sourcing workers can have prior experience with similar studies (i.e., they are non-naive) and bring their pre-assumed knowledge into the study, which in turn reduces the effect size of the study [Chandler et al. (2015)]. Workers might also discuss the experiment on the worker discussion boards where they might obtain foreknowledge about the experiment and be influenced by the opinions of other people [Chandler et al. (2014)]. Another HRI-specific concern is that online surveys and passive observation may not reflect participants' perceptions and interactions with the robot in real life. Guidelines on crowd-sourcing platforms can be found in [Thomas and Clifford (2017)]. Generally, demographic data on crowd-sourcing platforms cannot be verified [Harris (2011)], which can impact the studies with specific inclusion criteria. The participants may also be biased if there is a financial incentive [Eickhoff and de Vries (2013)].

4.3.2 Challenges in Recruitment

Researchers may face challenges in recruiting sufficient participants to take part in their study. Significant recruitment difficulties may indicate that the study design needs to be re-evaluated (e.g., due to the choice of task difficulty, lack of marketing material, and accessibility of the trial location). While study design may be a an underlying factor, some participants are inherently more difficult to recruit, such as those needed for long-term studies or studies involving specific groups (e.g., patients, children). For HRI studies, participants, such as patients, are often recruited through external collaborators. For instance, Lemaignan et al. (2016) recruited children with handwriting difficulties through the collaborating therapist. Once the participants who are more difficult to recruit have enrolled, researchers might inquire about the participant's future availability and willingness to partake in other relevant studies. However, researchers might also need to consider when not to reuse participants. Scenarios where participants should not be reused include when the study requires first exposure to the system. Participants who have prior experience in other studies need to be excluded to prevent cross-condition contamination [Mothershaw (2018)].

Regardless of the recruitment method, special attention is needed when recruiting vulnerable people. Vulnerability is not limited to disabilities but also includes instances where people might be exploited and abused due to an existing unequal or dependent relationship [Smith (2008)], for instance, students participating in their supervisor's study or clinicians recruiting patients for trials. This also includes recruiting close friends or family, who may comply with the experimental manipulation to please the researcher. A simple way to minimise the effect of potentially unequal relationships is through information disclosure (i.e., who is involved with the project, how privacy will be protected, and how patients' right to medical care will be reserved regardless of the study results).

Motivating participants to enroll in the study might be another challenge. Different incentives might be provided, such as food for students, gifts for children, raffles, and participant honoraria. Participant compensation may be proportional to the amount of time required and the type of participants involved. However, biases might be introduced when compensation is offered, as financial incentives might cause data from a subset of participants to be over-represented in the entire sample [Swanson and Betensky (2015)] and interfere with the participants' ability to weigh the risks and benefits of the study [Fry et al. (2006)]. When compensation is involved, the rule of thumb is that the payment should not be provided as a reward for any risk or harm [Fry et al. (2006)]. Researchers are to follow the standard procedure defined by the host research institute and maintain transparency. Ethics committees might only allow a small amount of incidental expenses to be reimbursed. In addition, researchers can motivate participants by providing accommodations, depending on the context of the study, such as offering to conduct the experiment at the participant's home to accommodate for those who might not be able to travel. Finally, it is important to design suitable incentives that do not cause additional risk for participation or unfairly encourage people who are not able to voluntarily consent to participate, for example, if they are financially challenged and participate in the trial only to receive the incentive.

In the context of participatory design studies, it is important for researchers to communicate upfront about how the intellectual inputs of the participants may benefit the study outcomes and establish suitable compensation or acknowledgement to such contributions [Kelly (2019)]. Ultimately, the transparency regarding the study itself, the risks and benefits, as well as respecting the participant's autonomy and privacy should be the foundation of an ethical HRI study design during the recruitment and incentivisation stage.

4.4 Participant Retention

Longitudinal studies are particularly susceptible to the problem of participant retention. Participants dropping out over time can result in a type of sampling bias known as selective attrition [Twisk and de Vente (2002)]. In clinical research, participants can drop out of the experiment if they do not feel the benefit of the therapy, causing the remaining sample to be biased towards the groups with more effective treatment [Patten and Newhart (2017)]. In the HRI context, selective attrition is observed when participants stop using a robot over time, especially in longitudinal, in-home studies [Fernaeus et al. (2010), de Graaf et al. (2016)]. The reasons for the participant's departure may not be random, and it is important to record selective attrition in longitudinal studies and its occurrence in each group. Clinical studies with high retention rates tend to have a "personalised" experiment design, for example, accommodating for different circumstances through flexible scheduling and providing transportation, and ongoing engagement throughout the study, for example, providing regular newsletters [Abshire et al. (2017)] and public seminars and fostering a community among participants.

4.5 Participant Preparation and Training

4.5.1 Preparation and Debriefing

Once the appropriate participants are recruited for the study, it is important to consider the need for participant preparation and training, especially in experiments that involve observers or operators. In the HRI literature, the details of the training program including its length and approach are often not reported [Riek (2012)]. The following questions can guide experimenters when planning for participant preparation and training. A full description of participant debriefing is available in Allen (2017).

- What information does the participant need prior to the start of the experiment to give informed consent? Is deception involved?
- What training does the participant need to undergo?
- How will the observers be trained? For example, what scale is being used to evaluate the HRI session, and what are the evaluation standards?
- Do the initial results present good inter-rater and intra-rater agreement? For instance, some experiments might require observers to undergo training so that they can label affects from people's facial expressions. Researchers can administer a test prior to the start of the actual experiment to determine whether the participants have been trained sufficiently. If the initial results show poor rater agreement, researchers might consider extending the training or removing some observers.

4.5.2 WoZ Operator Training

Beyond participants, in WoZ HRI studies, the WoZ operators (also known as "wizards") help the study design to bypass technical limits, such as in speech processing, manipulation, navigation, or localisation. Riek (2012) reviewed considerations of WoZ studies. Sufficient training of the wizards, signified by them reaching a specific speed and accuracy threshold, is required to ensure that the experiment proceeds smoothly and the wizard's operations are consistent between sessions.

During training, the capability of the wizard and interaction scope need to be fully defined and remain consistent across the experiments. This can be determined by setting specific speed and accuracy thresholds that the operators need to achieve before concluding the training. For example, if a participant asks a question to the wizard that is outside of the scope of the interaction, the wizard is to answer by mimicking how a machine would react (e.g., "I don't know"), instead of inserting additional information. If wizards are required to carry out challenging tasks (e.g., forced to cheat as part of the study design), they should also be trained so that they are sufficiently prepared as part of the risk management.

In addition, observer training is important for structured observational studies to ensure that observers develop an understanding of what events to be aware of [Wilson and Streatfield (1981)], such as defining what constitutes an interaction breakdown and how to make note of the event. Researchers may consider using a training video (e.g., recording a pilot study) to familiarise observers with the experimental procedure. Standards for training efficacy should also be established.

4.6 Privacy and Ethical Considerations

This section discusses other considerations about participants from the procedural and ethical aspects of the experiment, including privacy protection, informed consent, and special considerations for different experiment setups.

4.6.1 Privacy

Privacy is a core ethical principle in HRI studies because it protects participants from potential harms, including embarrassment or fear of being judged by others. For HRI studies that involve more sensitive data (e.g., health data), privacy violations can lead to loss of health insurance or employment or other detrimental consequences [Rothstein (2007)]. The basic guidelines noted by Lazar et al. (2017) include the following:

- Collect the minimum amount of information.
- Limit the number of times data is used.
- Aim for full information disclosure whenever possible.
- Provide a secure data storage medium (especially when cloud services are used).
- Remove personal identifiers whenever possible.
- Follow standard data disposal procedures recommended by the institution.
- Provide channels to address concerns and enforce accountability.

Beyond considering risks to participants' privacy caused by intentional data collection during research studies, unintentional data collected by a robot to perform its functions, such as video streaming for navigation or audio recording for speech-based activation, can have additional privacy implications, especially when the robot is deployed in a private setting such as a home environment. Privacy considerations need to be given especially to participants who experience memory impairments (e.g., Alzheimer's) [Sharkey and Sharkey (2012)]. In longitudinal HRI studies, when robots with sensing capabilities are deployed in home settings, for example, the PARO robot that has speech recognition and other capabilities to monitor and respond to the environment [Shibata et al. (2012), Moyle et al. (2017)], participants with memory impairments (and even those without impairments) may forget the presence of these sensors and think they are in the privacy of their own home. Furthermore, when a robot is deployed in a shared or public space, the privacy of bystanders and people who may have incidental encounters with the robot needs to be considered in the design [Rosenthal-von der Pütten et al. (2020), Salvini et al. (2010)]. In some cases, strategies such as adopting an "opt-in" approach, that is, avoiding preemptive choices [IEEE (2021)] and presenting a transparent, intuitive interface that allows user customisation based on an understanding of a robot's capabilities [Rueben et al. (2018)], may better respect the users' choices.

Another consideration is data storage during and after the completion of an HRI study. Storing the research data securely, privately, and temporarily, for instance, as protected files onboard the robot, may be a way to improve participant privacy. However, data sharing improves transparency and reproducibility of a study and avoids repeated collection of sensitive or expensive datasets, which helps not only the research community but also policymakers and the general public [Kowalczyk and Shankar (2011)]. To protect the participant's privacy while supporting open data sharing, individual researchers must derive data-cleaning procedures that properly anonymise or de-identify the data. For HRI studies, the data may include sensitive or personal information (e.g., names of participants, health conditions, faces of participants) in various formats (e.g., numeric data, text, video recordings). Different algorithms have been proposed to address data anonymisation, including Gaussian spatial blur, the k-anonymisation algorithm for removing sensitive attributes in a dataset [Kowalczyk and Shankar (2011)], and the Philter algorithm for text records [Norgeot et al. (2020)]. Beyond individual studies, institutional support can help protect participant privacy by creating a secure data-storage and sharing infrastructure and managing external user access [Kowalczyk and Shankar (2011)].

4.6.2 Consent

A recent review by Rosén et al. (2021) found that informed consent is the most common form of consent used in HRI studies. Despite the high adoption rate, 40% of the publications surveyed did not report what their consent process was. To obtain informed consent, HRI study participants must understand the purpose, procedure, benefits, and risks involved with their participation and their rights to data and information. As will be discussed in Section 8.3.3, researchers may be required to obtain extended consent from users to enable data sharing. However, there are scenarios where obtaining informed consent could be difficult. For experiment setups that require deception (e.g., WoZ studies), researchers have to purposefully withhold information at the start of the study. While using deception raises ethical concerns, it is sometimes necessary, especially when the phenomenon under investigation is prone to reactivity, that is, when the measurement changes the participant's behaviour [Price et al. (2015)]. It is important to devise a post-study debriefing procedure for this type of HRI study to fulfil the researchers' duty in information disclosure. A way to ensure basic ethical study design is by applying for ethics clearance with the affiliated institutions. Some research projects in collaboration with industry may choose to go through commercial ethics review boards. However, this is not recommended because commercial review boards might be subject to conflicts of interest [Lemmens and Freedman (2000)]. For specific HRI participant groups, such as children with autism, additional approval from expert reviewers and continuous supervision from governing or advisory boards may be required.

Consent may sometimes be implied or be unfeasible to obtain, such as when robots are deployed in public spaces, the public interacts with service robots, [Sarathy et al. (2019)], or the study includes specific participant groups (see Section 8.3.3). In public spaces, passive consent (e.g., putting up signs to inform the public that the area is being recorded) may be used to address privacy concerns. In the case of human interaction with service robots (e.g., a robot receptionist), implied consent is given when a customer volunteers their information in exchange for service. However, Riek and Howard (2014) and Sarathy et al. (2019) highlight concerns where a robot's consent-free behaviour could become behavior that requires consent. For instance, as a robot receptionist is recording a conversation with a customer, it may accidentally record a secondary conversation that the customer is having with another person, which could compromise the privacy of both individuals. Examples of accidental recording are seen in smart devices, where the devices constantly tap into private conversations in the absence of consent (e.g., Google Home recording audio clips despite the lack of the trigger phrase "OK Google") [Xia and Jiang (2020)]. Sarathy et al. (2019) proposed several research directions to begin addressing the dynamic nature of consent, such as disabling robot features until explicit consent is given (e.g., a Roomba not using its sensors until a user turns it on) or improving the ability of robots to understand social cues (e.g., infer from a user's gestures that its action violated consent).

4.6.3 Ethics Application

Most academic research projects operate under guidelines established by their governments or institutions. For instance, the most recent guidelines for European Union (EU) projects are defined in the Horizon 2020 framework, and the original principles are derived from the Charter of Fundamental Rights of the European Union and the European Convention on Human Rights. Projects in the United States are fundamentally based on the Belmont report and the Hippocratic Oath. In Canada, the Tri-Agency

Policies and Guidelines govern all ethical conduct for human research. In Australia, the standards are defined under the National Statement on Ethical Conduct in Human Research. While pilot studies involving only internal researchers may often begin during the development stage, formal user studies may only involve recruited participants after ethics approval has been granted.

In situations that cause ethical dilemmas or challenges in HRI studies, ethical guidelines and human research ethical approval guidelines can help in determining appropriate conduct. Examples of ethical dilemmas include participants expressing an interest in withdrawing from participation, asking for their data to be deleted, or raising concerns about the conduct of the trial. Researchers must prepare a research protocol and discuss with team members how the team will systematically approach challenges or ethical dilemmas before they arise, so the team is better equipped to deal with the situation appropriately as it happens.

4.6.4 Further Considerations

In addition to being aware of considerations for participants who might require special care, accommodation, or region-specific standards when preparing the experiment protocol, researchers should consider the potential concerns highlighted in this section relating to other special participant groups and research practices.

As HRI is a relatively new field, many phenomena are not well understood and researchers need to be aware of the potential long-term impacts on the participants across all age groups. For instance, Lemaignan et al. (2016) conducted a study with a social robot in a classroom setting. They pointed out the potential impact of the robot to change the dynamics between the teachers and the students in the classroom. In addition, de Graaf (2016) noted that long-term Human-Robot Interactions can potentially foster human-robot relationships, where humans start forming emotional bonds with the technology. Considerations are to be given beyond risk assessment from the safety perspective. The role and utility of the technology in the user's life need to be taken into account, especially in settings such as homes, long-term care facilities, workplaces, and schools. Researchers should be aware of the ethical implications involved when developing robots with human-like qualities or a high level of autonomy. These functions might fundamentally be a form of deception, as the users might wrongly believe that the robots have human characteristics when they do not.

Participants, such as WoZ operators, have the potential to have a negative experience during the study. For instance, Rea et al. (2017) reviewed social HRI experiments that put the WoZ operators in stressful situations, such as having to cheat [Short et al. (2010)], enduring verbal and/or physical abuse, or simulating other behaviours that would otherwise be socially awkward (e.g., long periods of silence). Thus, it is necessary to minimise the duration of the necessary social conflicts in the interaction script, allow the operators to have positive interactions with the participants in post-experiment debriefing, and have means for the operators to be trained and express their concerns [Rea et al. (2017)].

COVID-19 has highlighted the importance of hygiene in every procedure, including experimental studies, to protect both the participant and the researcher. Experimental procedures need to take into account the appropriate institution and government guidelines relating to social distance and minimise contact whenever possible. As some robots cannot be sanitised using alcohol-based methods, it is important to establish procedures between the host institution and the lab to ensure proper hygiene, for example, having people who need to work with the robots sanitise their hands before and after working with the robot or using near UV light for equipment sterilisation [Kitagawa et al. (2021)].

4.7 Examples

The study conducted by Moyle et al. (2017) demonstrates sample randomisation and stratification techniques. The authors investigated the use of PARO robots for treating the behavioural and psychological symptoms of dementia. The experiment took place in 28 different long-term care facilities, where the facilities were first stratified based on the organisation type (private vs. nonprofit) and then the experimental conditions were randomly assigned in blocks. A cluster randomised control trial was employed to minimise the effect of between-group contamination, as patients in different care facilities are inevitably exposed to different activities and treatment environments.

Lemaignan et al. (2016) highlighted the procedure for participant recruitment and study design in a study aiming to integrate robots into childhood education for both normal children and children with disabilities. This study demonstrated how an iterative study design can be used to address the challenges of validation with a vulnerable population group. The system was first validated with a more accessible population and then with the clinical group. Finally, the authors conducted two in-depth case studies at the lab with two children over a month. Since the study took place under multiple experimental settings (i.e., schools, the clinic, and the lab), the authors were able to identify differences in the children's behaviour in the lab compared to other settings.

The work described in Kanda et al. (2009) showcases participant recruitment in a target field environment. The authors released a semi-autonomous robot in a shopping mall, where it interacted with visitors every weekday for 4 hours over 5 weeks. During the first 3 weeks of the study, 332 participants were recruited by flyers that were distributed around the mall or approached by the experimenters who were on-site. At the end of the study, a questionnaire was mailed to each participant and 70% of the participants responded.

Wang et al. (2010) illustrates the importance of considering cultural context during recruitment. The study explored how the cultural background of the participants (Chinese vs. American) affects the human-robot collaboration process when using two different communication styles (implicit vs. explicit). The research team was able to find representative samples by recruiting participants directly from two comparable universities in China and the United States and conducting the experiments in the respective countries.

Yanco et al. (2015) provides an example showing the recruitment of specialised participant groups. The study evaluated in-situ HRI interfaces developed for expert users for use during the DARPA Robotics Challenge. The authors used the unique event to collect data with expert users. They observed the teams participating during the trials and analysed the HRI methodologies used for robot teleoperation.

4.8 Summary

We discussed the types of HRI study participants and considerations for managing the benefits and risks to them. Equality, inclusiveness, and fairness of an HRI study can be improved by recruiting a diverse group of participants who best represent stakeholders of the HRI system. Recruiting representative participants is also key to validating that the resulting findings and HRI system designs align with the prioritised ethical values and requirements of the stakeholders. Within the context of collaborative design and evaluation of an HRI system, current compensation norms may not sufficiently reward participants' contributions of their time and intellectual insights. To develop HRI systems that respect human agency and autonomy, beyond the informed consent and ethics approval practices already in place, HRI research can benefit from transparent and continuous mutual communication with the stakeholders regarding study design, privacy and data management, and expected outcomes in the short and long term.

Recommendations for Selecting and Recruiting Participants for HRI Studies

Participant selection:

1. Use unbiased/random sampling when a phenomenon relating to a population is under investigation. However, unbiased sampling can be difficult to achieve due to practical limitations (e.g., participant accessibility, availability, and proximity) and convenience or purposive sampling is often used instead. Balancing potential confounding variables in each participant group (e.g., gender or educational background) can help with noise introduced by non-random sampling techniques.
2. Use simple randomisation, such as a coin toss or random number generator, when there are a large number of participants, but do not use it for a small participant group. Stratified randomisation is best used for equalising the number of participants across multiple factors, such as age. Block randomisation and response adaptive randomisation are more sophisticated techniques.
3. Report blinding in the final report because it represents an important methodological step. Allocation concealment is an important step for demonstrating that the experiment is not influenced by the researcher's bias.

Recruitment and incentivisation: Each recruitment platform comes with different advantages and trade-offs (see Table 4.1). In addition, suitable incentivisation can be considered, but take care that it does not affect the participant's consent.

Retention: Participant retention is important, especially in longitudinal studies. Include the participant drop-out rate in the final study report.

Training: Ensure that the participants are sufficiently trained. Describe participant preparation, such as training for WoZ operators/teleoperators, in the study.

Privacy and ethics:

1. Maintain participant privacy through data anonymisation and data management.
2. Informed consent may not be possible for WoZ studies. Appropriate procedures need to be put in place to debrief the participants after the study.
3. Given that HRI is an emerging field, there may be long-term consequences of the experience that are unknown. Considerations need to be given to all possible participants, including study participants who were recruited, such as children or vulnerable groups, as well as researchers and designers who were involved in the experiments.

5

Quantitative Evaluation Methods

In this chapter, we focus on quantitative evaluation methods for HRI experiments. We review common constructs and measurements for HRI experiments designed to verify hypotheses in confirmatory HRI studies. For qualitative and mixed methods, see Chapter 11.

5.1 Common Constructs in HRI Studies

The choice of constructs (what to measure) and metrics (how to measure it) is important, given that they will determine the focus of the research and the data to collect. The constructs and metrics will help assess the validity of the research hypothesis and the relevance of the variables controlled by the researchers, but also shed insights on the implications for the other agents involved in the interaction process or the interaction itself. We provide a summary of the common constructs and their related metrics in Section 5.2.6 Table 5.2. Researchers are encouraged to evaluate the quality and soundness of the chosen measurements because they affect the statistical analysis of the study. This can be done by evaluating the statistics used to report the face, content, criterion, and concurrence validity of the measurement.

HRI studies are often interested in validating theoretical concepts, known as constructs. Most of these constructs are not directly observable but can be linked to observable and/or measurable metrics and events [Hyland (1981)]. Given the wide range of applications in HRI, more than 110 different metrics have been proposed or used in the HRI literature [Damacharla et al. (2018)]. These metrics can be grouped by the agents participating or measuring during the interaction process, which includes the robot, the human, and the team (often referred to as the system) [Murphy and Schreckenghost (2013)]. Given the wide scope of HRI studies, the metrics listed in this section might not apply to all types of user studies.

- *Robot-related*: This metric includes the level of robot autonomy [e.g., Olsen and Goodrich (2003) and Beer et al. (2014)] and the evaluation of a robot's

DOI: 10.1201/9781032693682-6

social attributes and people's intention to use it [e.g., Carpinella et al. (2017) and Robinson et al. (2018)].

- *Human-related*: This metric includes training outcomes, human learning performance, and adapting to the task [e.g., Harmon-Jones et al. (2016) and Anzalone et al. (2015)].

- *System-related*: This metric includes the team dynamics, roles of each member, and task outcomes [e.g., Hoffman (2019b) and Damacharla et al. (2018)].

Although quantitative methods are often used to validate functional requirements, they can also be employed to analyse ethical requirements. Task performance metrics can assess how HRI can benefit people by improving work efficiency, such as shorter task completion times or a higher task success rate. These task measures can also assess the feasibility and suitability of robot use in the proposed task or use case. This includes determining to what extent the number of safety-related events or the mean rating of intended uptake rates could impact the practical uptake of robots in particular domains. When using quantitative methods to evaluate these factors and outcomes, the reliability of the metric (the degree to which the metric is free from measurement errors) and validity (the degree to which the data support the theoretical properties they are supposed to measure) should be considered [Koo and Li (2016), Gergle and Tan (2014)].

5.1.1 Robot-Related Constructs

Robot-related constructs can help evaluate the attributes in the hypothesis relating to the level and behaviour of autonomy as defined in Section 2.2.1. Neglect tolerance [Olsen and Goodrich (2003)] (i.e., how the robot effectiveness declines when the human is not attending the robot) and attention demand (the percentage of time the human must control the robot) are often used as general measures of a robot's *autonomy*. Beer et al. (2014) suggested that a subjective rating of human intervention as a supplementary measure of the level of autonomy of a robot also be included.

Productivity is another construct that is related to the robot's performance at a given task and/or the robot's technical capabilities (e.g., accuracy at object recognition). Time-based and error metrics, such as task completion time and number of unsuccessful actions, are often used to assess the robot's task performance [Olsen and Goodrich (2003)]. Finally, the third construct related to robots is the *social attributes*, which aims to capture how traits (e.g., warmth and competence) and characteristics (e.g., anthropomorphic appearance) associated with robots affect the social perception of people interacting with them. Since these attributes mostly rely on people's judgements, subjective scales such as the Robotic Social Attributes Scale (RoSAS) [Carpinella et al. (2017)], Robot Incentives or

Robot Self-Efficacy Scale [Robinson et al. (2018, 2020)], and Godspeed questionnaires [Bartneck et al. (2009)] are often employed as metrics.

5.1.2 Human-Related Constructs

Human-related constructs are related to the attributes of the hypothesis that involve adaptation, learning, and training, as well as the experimental task. For example, situation awareness and cognitive workload are two intertwined constructs that have been identified as particularly relevant in both automation and robotics [Steinfeld et al. (2006), Beer et al. (2014)]. *Situation awareness (SA)* defines the individual's perception and comprehension of the environment and the interactions of the elements within the environment (in simple terms, understanding "what's going on") [Salmon et al. (2009)]. *Cognitive workload (CW)* is a product of the mental resources required by a task and the capacity of the person performing the task. Both SA and CW can be important in teleoperation tasks, such as in search and rescue missions. When robots are deployed in a disaster area, integrating multiple sensing capabilities (e.g., depth and proprioceptive feedback) with reduced time delay can allow the operator to develop a better SA [Murphy and Burke (2005)]. The operator's CW can also be managed with strategic information display (e.g., when to raise an alert), which can ultimately reduce stress and facilitate decision making [Prewett et al. (2010)]. In addition, several other constructs have been used to investigate people's perception of robots, their responses to different robots under different contexts, and types of interaction [Bartneck et al. (2020b)], or even to predict future use of the robot [Beer et al. (2014)]. For example, constructs such as *willingness to interact with the robot* [Heerink et al. (2009), Robinson et al. (2018)], *trust*, *perceived or physiological safety*, *engagement*, and *affective states* have been shown to be potential predictors of robot use [Beer et al. (2014)]. Measurement or assessment of these constructs is often done using questionnaires, such as the RoSAS [Carpinella et al. (2017)] or Discrete Emotions Questionnaire [Harmon-Jones et al. (2016)] or by physiological measurements, such as changes in heart rate and skin conductivity [Anzalone et al. (2015)] and behavioral measurements.

5.1.3 System-Related Constructs

System-related constructs are related to the team structure and task attributes of the hypothesis. Most of the constructs associated with the system are defined in the context of task-oriented applications, which aim to assess how well the human(s) and robot(s) perform as a team. For instance, while *productivity* and *efficiency* [Murphy and Schreckenghost (2013)] are associated with how well the task is completed and the time and effort required to complete the task, *fluency* characterises the coordination of joint activities and actions between members of a well-synchronised team

[Hoffman (2019b), Hoffman and Breazeal (2007)]. Metrics such as the *time elapsed* between the end of an agent's action and the beginning of the other agent's action (i.e., the functional delay), *number of unplanned human interventions or interactions*, and the *time required to interact with the robot* (i.e., the interaction effort) are often used to measure how different experimental conditions, human factors, or new robot skills affect the system's performance [Damacharla et al. (2018)].

Recommendation: The various constructs mentioned in this section attempt to break down the interaction in a systematic way and provide metrics that can be used to quantify the interaction. We recommend reviewing Table 5.2 in Section 5.2.6 a quick overview of the common constructs and measurements to select what is best for each trial.

5.2 Common Quantitative Measurements in HRI Studies

Multiple constructs can be of interest in an HRI study. However, existing studies often attempt to capture the HRI experience with multiple constructs using a single measurement, which is generally insufficient [Bethel et al. (2020b)]. Choosing the right measurements often requires specific expertise in the phenomenon under investigation to bridge the knowledge gap in an interdisciplinary HRI study. For example, facial expression alone might not be the best indicator for evaluating emotions, as expressions can be culture-dependent or displayed outside the context (e.g., adults learn to smile to hide disappointment or awkwardness) [Barrett et al. (2019)]. This section highlights common evaluation methods listed in Bethel et al. (2020b) and discusses the advantages, disadvantages, and application of each. Note that here we have focused on quantitative measurements. Qualitative measures, such as interviews, can also be powerful in HRI evaluation. We will discuss qualitative and mixed-method evaluation in Chapter 11.

5.2.1 Task Performance Metrics

Task performance metrics are objective measurements of how well the robot, the participant, and the overall system perform the task under investigation. Performance can be measured with respect to five different aspects of the robot's capability, which are navigation, perception, management, manipulation, and sociability [Steinfeld et al. (2006), Bethel et al. (2020b)]. Examples of these metrics include task completion time, error rate, and efficiency. More examples can be found in Table 5.2 in Section 5.2.6.

With respect to the human, task performance can be measured as subjective experiences or objective outcomes of the person, with metrics such as the person's productivity or situation awareness [Salmon et al. (2009)]. On the interaction level, task performance can be measured as the flow and dynamic of the human-robot system, with metrics such as fluency [Hoffman (2019b)] or coactivity [Murphy and Schreckenghost (2013)]. For a comprehensive review of task performance metrics, see Bethel et al. (2020b) and Rueben et al. (2020a).

Performance measurements are useful for comparing technologies, as they give a numerical value indicating how much improvement can be achieved by adopting the technology (e.g., introducing warehouse robots to improve the sorting efficiency of packages). However, task performance metrics alone are insufficient to capture the entire HRI experience, and the result might be heavily influenced by the population. For example, people in engineering schools with more exposure to a range of technologies might have a shorter task completion time on a human-robot collaboration task compared to people who are less experienced with technology. The effect of population can be reduced by selecting a large, diverse group of participants and introducing a control/baseline to measure relative changes instead of the absolute values.

Recommendation: Task performance metrics can be used to measure the performance of the robot, the participant, and the benefit of the interaction. The metrics should not be used alone to substantiate claims about the entire HRI experience.

5.2.2 Behavioural Measurements

Humans are often involved in the evaluation process to provide direct feedback on the proposed robot use case or their perception of the robot's performance, such as using self-reported surveys or an observer's ratings.

Quantitative measurement can also be derived from data collected during the interaction, such as linguistic analysis [Fischer et al. (2013)] or field notes. For instance, Fischer et al. (2013) hypothesised that introducing contingent (i.e., temporal, non-accidental actions from a robot in response to a user's inputs) verbal feedback in a robot may improve its perceived sociability. The authors compared the impact of contingent behaviour using metrics including verbosity (i.e., the number of words and turns taken during HRI), number of attention-getting events (e.g., by issuing commands to the robot or referring to the robot's name), word diversity (i.e., the number of different words used), and sentence complexity (i.e., the average utterance length). Collectively, human-centred measures can provide greater insight into whether people report that the robot intervention has taken into consideration their needs and requirements. Behavioural data can be recorded under four different study contexts:

- Naturalistic observation involves observing people's behaviour in a natural environment where it typically happens (quantitative or qualitative).
- Participant observation is where researchers are part of the experiment they are conducting (quantitative or qualitative).

- Structured observation, where researchers only focus on gathering quantitative data of the behaviours under investigation in a particular experimental setup, are typically more efficient than naturalistic or participant observations because researchers will already know what to look for during the experiment.
- Case studies are a detailed investigation of an individual, social unit, or event [Price et al. (2015)].

A metric in HRI studies might include recording how long participants look at the robot. Behavioural measurements do not rely on the recollection of the participants or the self-interpretation of their behaviour [Bethel et al. (2020b)]. However, behavioural measurements are susceptible to the Hawthorne effect, where participants might behave differently from how they normally would because they know that they are being watched. A way to obtain more accurate behavioural measurements is to combine multiple signal-level behavioural metrics. For example, measuring changes in pen pressure and stroke length along with handwriting could help distinguish whether participants are being truthful in their responses [Luria and Rosenblum (2010)]. Behavioural data can also be obtained in the form of audio or videotape, requires significant time and expertise to analyse. The cost of data annotation is often high, and many researchers might not process the data. With annotation comes the problem of inter-rater and intra-rater variability. Finally, behavioural measurements alone are insufficient to explain participant behaviour [Johnson et al. (2016)] and are better used in conjunction with other measurements, such as interviews (see Section 11.1.2).

Furthermore, behavioural data can be obtained via specialised sensors for understanding a participant's cognitive, affective, and physical states. For example, eye gaze data can be used for analysing intentions during human-robot handovers [Moon et al. (2014)] or understanding an operator's situational awareness when monitoring a remote robot fleet [Senaratne et al. (2023)]. The advancements in wearable sensors also facilitates the use of physiological data in HRI research, such as muscle activation and heart rate, especially for understanding physical interaction, assistive robotics, and exoskeletons [Wu et al. (2021), Heinisch et al. (2024)]. In Section 5.2.3, we will further discuss the use of psychophysiological data as a quantitative measurement in HRI experimental studies.

Recommendation: Behavioural measurements quantify human-related constructs without the need to rely on the participant. When using these measurements, psychologic effects must be taken into account in data interpretation and reporting of the final results. Sufficient training should be given to the researchers who are rating the interaction to address inter-rater and intra-rater variability. Behavioural measurements can be used in conjunction with other metrics.

5.2.3 Psychophysiological Measurements

Psychophysiological measurements are physiological signals affected by a participant's mental state, which can provide additional information to standard quantitative questionnaires on how people perceive and interpret the presence or assistance of robots. Example measurements include heart rate, blood pressure, brain activity, skin conductance, and muscle activity [Bethel et al. (2007)]. These measurements are good for quantifying abstract concepts such as physiological arousal using non-invasive and objective measurements. The time domain aspect of these measurements helps to pinpoint exactly when a response was elicited in the experiment [Dautenhahn and Saunders (2011)]. Furthermore, an advantage of using psychophysiological measurements compared to behavioural measurements is that participants cannot easily manipulate an automatic response. For a detailed review of the psychophysiological measurements commonly used in HRI, see Bethel et al. (2007).

Psychophysiological measurements are informative when applied properly. Since the signal gives very specific information about the person's state (e.g., heart rate and respiration rate are indicators for arousal levels), it is important to understand what the signal is measuring and not to infer meanings that may not be accurate. One challenge of obtaining the measurement is that the signal is often influenced by various confounding variables and noise (e.g., health status and room temperature). Health conditions can prevent changes with respect to the baseline to be observed due to the Law of Initial Values. The law states that the measurement can increase or decrease within a narrower range if the initial measurement is already high or low [Bethel et al. (2007)]. Other confounding factors include the orientation response, the defensive response, the startle response, and habituation [Bethel et al. (2007)]. For example, a loud noise from the robot during the experiment could make the participant nervous, resulting in a temporary increase in heart rate that is not related to the actual experiment. There may be scenarios in which the measurement does not provide useful information, for example, if the task might not provide a significant change from baseline or when different tasks might produce the same psychophysiological readings [Dautenhahn and Saunders (2011)].

Recommendation: Psychophysiological measurements provide detailed information on the participant's internal state without relying on self-reported measures. These measurements should be used with caution. They require significant expertise to ensure the correct measurements are taken and the right interpretations and conclusions are drawn. Researchers should complete the relevant training and read manuals on how a specific psychophysiological signal is measured before using it in the test.

5.2.4 Self-Assessment

Self-assessments in HRI studies can be both quantitative and qualitative and often come in the form of paper- or computer-based scales, questionnaires, or surveys. Examples include the GodSpeed questionnaire, which measures human perception of a robot's anthropomorphism, animacy, likeability, perceived intelligence, and perceived safety [Bartneck et al. (2009)], and the stress reported by operators in a collaborative assembly of humans and robots [Arai et al. (2010)]. A list of commonly used surveys can be found in Table 5.2 in Section 5.2.6.

Assessments are easy to administer but are influenced by many factors and limited by human performance. A factor that affects the accuracy of the measurement is the timing of the survey. Since surveys are often administered at the end of the study, it means that data can only be collected on a reflective level, researchers will not be able to corroborate the information from the participant directly, and participants will need to remember the feelings [Dautenhahn and Saunders (2011), Bethel et al. (2007)]. Assessment is also influenced by social or cultural norms, where participants might answer questions based on what they think is acceptable by the research team or society [Bethel et al. (2007)]. Some self-assessment measures require substantial fees to be paid for their use, increasing the cost of research. Others have copyright that must be approved by the owner for its use, which might be difficult to obtain.

Recommendation: Self-assessment data are often collected through the use of surveys and questionnaires. Self-assessments can also be influenced by various human factors, such as how the researcher introduces the use of the questionnaire or the instructions given on its completion. Whenever possible, it is better to use validated existing questionnaires.

5.2.5 Questionnaire and Survey Design

Existing surveys may not always be sufficient to capture the variable at hand or the research direction of interest. Researchers can consider designing their own questionnaire, but proper validation requires extensive testing, statistical measurement, and sampling. Adopting existing questionnaires is recommended if they have been validated. Questionnaires cannot be altered once validated, as they can become invalid when the order of the items or wording is changed [Rust and Golombok (2014)].

To create a new questionnaire, the first step is to select the appropriate measurement scale (i.e., nominal data, ordinal data, interval data, or ratio; see Table 5.1 for more information). Once the scale is selected, the types of options and the number of options need to be considered.

TABLE 5.1
Choice of Measurement Scales

Value Type	Definition and Example
Nominal data	Categorised data (e.g., male participants are assigned the value of 1 and females are assigned the value of 2)
Ordinal data	Ranked data (e.g., participants ranking their preferences between robots on a discrete scale from 1 to 10)
Interval data	Data measured along a scale (e.g., participants rating their enjoyment level on a continuous scale from 0 to 100 or the participant's objective score on a math test from 0 to 100)
Ratio	Ratio between two measurements (e.g., comparing the enjoyment level of participants in two groups by using the ratio of the interval data collected from the two groups)

Rust and Golombok (2014) provide a general guideline for the design of question options:

- A personality or mood questionnaire often includes options such as "not at all", "somewhat", and "very much".
- An attitude questionnaire might have "strongly disagree", "disagree", "neutral", "agree", and "strongly agree" as options.
- A clinical symptom evaluation scale might include "always", "sometimes", "occasionally", "hardly ever", and "never" as options.

The number of options for each questionnaire item depends on the nature of the questionnaire. The important factor is to provide enough options to participants so that they can express their opinions. A general rule is to include at least four options for rating scales and to use a consistent number and type of options throughout the survey [Rust and Golombok (2014)]. When possible, test different questionnaire wording and choice of variables using a pilot study to improve the quality of the survey. Johnson et al. (2016) listed a detailed procedure to construct a questionnaire in educational research, and the same principles can be adopted in HRI studies. In general, it is important for researchers to note that customised questions have not yet been tested outside the experiment, and when feasible, provide evaluation metrics for validating the survey (see Section 5.3.1). For more detailed information on how to design a questionnaire, researchers may refer to Rust and Golombok (2014) and Litwin and Fink (2003). Examples in which customised questionnaires are designed and validated for HRI experiments can be found in Hoffman and Breazeal (2007) and Dragan et al. (2015).

Recommendation: The design of a questionnaire involves selecting what to measure (i.e., the actual questions) and how to measure them (e.g., the measurement scale, number of options, and phrasing of the questions). Psychometric validation is lengthy, complex, and time consuming. Researchers carefully consider the need to create a new scale and investigate other literature for suitability, such as scales in the human-computer interaction domain. If the scale is being used for the first time, this is reported as part of the measures section, advising caution with the interpretation. Reports on survey validity and reliability should be included when reporting the scale in a research trial.

5.2.6 Overview of Common Constructs

TABLE 5.2
Constructs and Related Metrics

Construct	Measurement	Metrics
	Robot	
Autonomy [Steinfeld et al. (2006)]	Task-performance metrics	Neglect tolerance and attention demand [Olsen and Goodrich (2003)]
Social Attributes [Carpinella et al. (2017)]	Self-reporting questionnaires	RoSAS [Carpinella et al. (2017)]
Productivity [Steinfeld et al. (2006)]	Task-performance metrics	Task completion time, percentage of successful actions [Steinfeld et al. (2006)]
	Human	
Situation awareness [Scholtz (2003)]	Task-performance metrics, self-report questionnaires, physiological measurements	Situation Awareness Global Assessment Technique, Situation Awareness Rating Technique [Endsley et al. (1998)], gaze analysis [Dini et al. (2017)], and secondary task performance [Salmon et al. (2009)]

Continued on next page

TABLE 5.2
Constructs and Related Metrics (Continued)

Construct	Measurement	Metrics
Cognitive workload [Steinfeld et al. (2006)]	Self-reporting questionnaires, task-performance metrics, physiological measurements	NASA-Task Load Index (TLX), Social Avoidance and Distress Scale (SADS) [Marvel et al. (2020)], task errors and reaction time [Prewett et al. (2010)], respiratory rate, heart rate, skin conductance, and temperature [Novak et al. (2011)]
Acceptance [Heerink et al. (2009)]	Self-reporting questionnaires	Godspeed questionnaires [Bartneck et al. (2009)], Technology Acceptance Model based questionnaires [Heerink et al. (2009)], Almere Model [Marvel et al. (2020)], Complacency Potential Rating Scale [Natarajan and Gombolay (2020)]
Trust [Steinfeld et al. (2006)]	Self-report questionanires, task-performance metrics	Human-Robot trust scale [Schaefer (2013)], trust in automation scale [Jian et al. (2000)], task allocation [Roncone et al. (2017)], Almere Model, BEHAVE-II, Multi-Dimensional Measure of Trust, and Negative Attitude toward Robots Scale (NARS) [Marvel et al. (2020)]
Affective state [Kulić and Croft (2007)]	Self-report questionnaires, physiological measurements	Discrete Emotions Questionnaire [Harmon-Jones et al. (2016)], Pleasure-Arousal-Dominance ratings [Betella and Verschure (2016)], and cardiovascular and electromyogram (EMG) activities [Kulić and Croft (2007)]

Continued on next page

TABLE 5.2
Constructs and Related Metrics (Continued)

Construct	Measurement	Metrics
Stress [Lasota et al. (2017)]	Self-report questionnaires, physiological measurements	Skin potential response, semantic differential (SD) questionnaire [Arai et al. (2010)], NASA-TLX, NARS, and SADS [Marvel et al. (2020)]
Perceived safety [Lasota et al. (2017)]	Self-report questionnaires, physiological and behavioral measurements	Godspeed (safety questionnaire) [Bartneck et al. (2009)], NARS [Nomura et al. (2006)], distance between human and robot [Mumm and Mutlu (2011)], and cardiovascular and electrodermal activity [Lasota et al. (2017)]
Engagement [Steinfeld et al. (2006)]	Physiological and behavioural measurements, self-assessment	Changes in heart rate and skin conductivity changes, non-verbal gestures [Anzalone et al. (2015)], Self Assessment Manikin Instrument [Marvel et al. (2020)]
	System	
Productivity [Steinfeld et al. (2006)]	Task-performance metrics, self-report questionnaires	Fan out and interaction effort [Olsen and Goodrich (2003)], System Usability Scale (SUS), Multidimensional Robot Attitude Scale (MRAS) [Marvel et al. (2020)], and productivity time [Damacharla et al. (2018)]
Fluidity [Steinfeld et al. (2006)]	Task-performance metrics	Agent idle time, concurrent activity, and functional delay [Hoffman (2019b)]
Efficiency [Murphy and Schreckenghost (2013)]	Task-performance metrics, self-reporting questionnaires	Human-robot action or time ratio, time to complete the task [Steinfeld et al. (2006)], perceived quality of interaction [Baraglia et al. (2017)], MRAS, and SUS [Marvel et al. (2020)]

Continued on next page

TABLE 5.2
Constructs and Related Metrics (Continued)

Construct	Measurement	Metrics
Reliability [Steinfeld et al. (2006)]	Task-performance metrics	False alarms and number of interventions [Murphy and Schreckenghost (2013)]

5.3 Other Considerations

It is important to consider the quality of a metric, which is the extent to which it can accurately (i.e., validity) and consistently (i.e., reliability) capture the construct of interest. Donmez et al. (2009) identified several other factors that need to be evaluated when selecting the type of measurements and the specific metrics that can be included in an HRI user study. These factors are:

a) *Experimental constraints*: These are constraints related to the study context and how a measurement is collected, such as the temporal and monetary resources associated with collecting and analysing a specific metric, and the characteristics of the testing environment (e.g., gaze-based metrics make sense in controlled, in-the-lab settings but are difficult to obtain in a field context).

b) *Comprehensive understanding*: This is the consideration on how much each selected metric explains the hypothesis of interest and the amount of additional insight that can be obtained from the casual relationship between metrics (e.g., a decrease in task performance can be explained by an increase in the human-partner mental workload).

c) *Statistical efficiency*: The selected metric needs to provide the researcher with enough measurements that the requirements for the statistical power to detect potential effects are met.

d) *Measurement efficiency*: The type of measurement used to collect a specific metric should not be intrusive or distracting to the participants and to the nature of the study, task, and interaction, unless specifically required.

5.3.1 Reliability

Reliability is concerned with the degree to which the metrics are free of measurement error. Formally, it is defined as the proportion of variance of a metric that is attributed to the true variance of the score (i.e., the metric that would be obtained if there were no measurement errors) [Suter (2011)]. Since true score variance cannot be calculated directly [Heale and Twycross (2015)], in practice, reliability is often described in terms of consistency over four potential sources of error (time, forms, items, and raters) and involves some type of correlation computation.

Consistency over time is often referred to as *test-retest reliability* and can be measured as the correlation coefficient between the metric values obtained under similar circumstances at two different moments in time (stability). Consistency between forms is known as *reliability of parallel form* and can be obtained by measuring the correlation between the metrics collected using a different form of the original instrument in an immediate retest procedure. Consistency across items, also known as *internal consistency reliability*, is the most commonly reported type of reliability and corresponds to the correlation between people's responses across the items on a multiple-item instrument. Cronbach's alpha is the most commonly used test to determine the internal consistency of an instrument [Heale and Twycross (2015)].

When raters are used to generate information that serves as a metric of interest, *inter-rater reliability* can be applied to determine the extent to which different observers are consistent in their judgements. Inter-rater reliability coefficients are sensitive to consistency in terms of relative agreement but do not determine absolute agreement (e.g., two observers agree on the relative rank ordering of their scores of a specific metric but attribute different values) [Suter (2011)]. An inter-rater reliability coefficient greater than 0.6 is considered satisfactory [Eliasziw et al. (1994)]. A kappa score can also be used to test inter-rater reliability [refer to McHugh (2012) for more information].

In addition to the inter-rater reliability, *intra-rater reliability* measures the consistency of one observer's ratings over time and is also calculated using a correlation coefficient. Researchers can capture this measurement by asking a rater to complete the same rating scale at two different time points. The baseline correlation coefficient typically ranges from 0.6 to 0.8, and when the calculated coefficient is greater than the baseline, it is considered to be in good agreement [Litwin and Fink (2003), Eliasziw et al. (1994)].

Independent of reliability type, it is important to note that reliability cannot be used to describe the measurement instrument (i.e., it is not the instrument that is reliable) [Suter (2011)]. Similarly, reliability scores are dependent on the participant sample being measured and, hence, need to be included when reporting a study's results.

Recommendation: Metric reliability captures the quality of data with respect to measurement errors. Test-retest reliability, internal consistency reliability, inter-rater reliability, and intra-rater reliability are the common measures that are applicable to HRI studies. If they can be quantified in the trial data, they are reported.

5.3.2 Validity

Validity is considered the most important quality of a measured dependent variable. It is defined as the degree to which empirical evidence, that is, the scales, metrics, and instruments employed in a study, actually measure and support the theoretical properties and constructs they are supposed to measure [Gergle and Tan (2014)]. The general recommendation regarding construct validity is to select metrics that have been validated when they are applicable [Deonmez et al. (2009)], for example, using the Affinity for Technology Interaction (ATI) scale [Franke et al. (2019)] to measure interaction styles. In cases where non-validated metrics are chosen, construct validity and the degree to which valid inferences can be made based on the proposed metrics are assessed by considering the following four aspects [Price et al. (2015), Ralph and Tempero (2018)]:

- *Face validity* is the extent to which a metric or measurement method appears to be intuitively reasonable and represents what the researcher is attempting to measure. Since face validity is based on people's intuitions about human behavior, it is considered to be a very weak evidence of construct validity [Gergle and Tan (2014)].

- *Content validity* is the extent to which a measurement reflects and covers all the content that it presumably samples. In other words, does the selected measurement encompass all aspects of the construct it was designed to measure [Heale and Twycross (2015)]? Content validity is often assessed using expert judgements.

- *Criterion validity* is the extent to which different measurements (also referred to as criteria) reflect the same construct and is measured in three ways [Heale and Twycross (2015)]: *convergent validity*, which shows that the metric of interest is highly correlated with other criteria that measure the same variable or construct; *concurrent validity*, which indicates the extent to which the measurement or metric being evaluated is related to other criteria or some other construct measured at the same moment in time; and *predictive validity*, which assesses the degree to which the metric of interest predicts things it theoretically ought to predict.

- *Discriminant validity* is the degree to which the metric being evaluated is not correlated with measures of variables that are conceptually distinct.

Recommendation: Validity is the extent to which the data-gathering tool measures the intended measurement. Face validity, content validity, criterion validity, and discriminant validity are four important metrics to consider when exploring metric validity in the HRI setting. If known, these are reported as part of the measure description.

5.4 Examples

An example of an HRI study focusing on human-related constructs (i.e., trust and acceptance) can be found in Natarajan and Gombolay (2020). The authors examined how trust and acceptance of the robot can be affected by the level of anthropomorphism, the robot's personality expressed through its verbal feedback, and physical presence (physical vs. virtual). Participants in the experiment had to complete a math quiz with the hints provided by the robot. The authors used a mixed-model factorial design and measured trust and acceptance using both self-assessment surveys (i.e., the Complacency Potential Rating Scale and Trust in Automated Systems Scale) and task performance metrics (i.e., the number of times participants accept low-quality advice or reject high-quality advice). The study demonstrates how multiple constructs can be measured simultaneously using multiple measurements.

Regarding system-related constructs, Hoffman and Breazeal (2007) pioneered the first fluency evaluation in HRI with a physical robot. The authors hypothesised that the human-robot collaboration fluency can be improved by replicating the anticipatory behaviour in human-human collaborations. The authors implemented a cognitive framework in a non-anthropomorphic robot and designed an experiment to evaluate the framework. As part of the experiment, the authors developed and evaluated a customised survey using proper techniques as described in Section 5.3.1.

5.5 Summary

Quantitative methods provide an objective assessment of how specific factors in HRI influence a population. On the scale of a single HRI study, it can be difficult to adopt multiple evaluation methods at once, and researchers are encouraged to choose the most suitable methods according to their research questions, study feasibility, and strengths and weaknesses of each method.

Task performance metrics, human evaluation metrics, and psychophysiological measurements each have limitations. Task performance and human evaluation measurements can be highly influenced by the population [Horstmann and Krämer (2019)]. For instance, participants with higher technical efficacy may have a shorter task completion time in a human-robot collaboration task, compared to those less experienced with technology. Careful data analysis or intentional participant screening at the beginning of a study (e.g., measuring participants technical efficacy or past experience using or interacting with robots) may help highlight inequality created by access to technology. The weakness of using psychophysiological signals is that they can be difficult to collect and interpret in a non-invasive way. These signals are often influenced by various confounding variables and noise (e.g., health status, room temperature, and habituation) and may not be reflective of the mental state being investigated [Bethel et al. (2007), Dautenhahn and Saunders (2011)]. For example, a sudden loud noise during an experiment could startle a participant, resulting in a temporary increase in heart rate that is unrelated to the HRI factors being investigated.

Moreover, hypothesis testing with quantitative methods assumes objectivity, with the HRI experience captured through numerical data points. The experiment paradigm also requires a sufficient sample size to produce statistically significant or generalisable results [Veling and McGinn (2021)]. This includes ensuring that a wide enough sample is collected to create fair and generalised assumptions about the targeted population before deploying the robot interaction. For example, samples should adequately represent different genders, technical skill levels, and socioeconomic backgrounds. Quantitative measurements alone may be insufficient to capture the entire HRI experience and, therefore, cannot be relied on solely to make clear recommendations that ethical design requirements have been met. As Seibt et al. (2021) and Veling and McGinn (2021) argued, HRI studies are "context-dependent" and "perspectival" and cannot be reduced to a quantitative measurement alone. Further, Seibt et al. (2021) pointed to the fact that because HRI is a relatively new field, the general public may still be developing social norms to accommodate and interact with the new robotic agent. Therefore, quantitative data may be biased by a novelty effect, as well as by participants' access and willingness to partake in the experiment, which limits the generalisability of the experimental result. Further discussion on biases can be found in Section 10.1.

Quantitative methods for assessing task results or immediate changes in users' mental states align with utilitarian ethical values, as will be discussed in Section 8.3. Beyond the aforementioned metrics, new quantitative metrics that focus on the virtue ethics and duty ethics, such as fairness [Bellamy et al. (2018)] and social acceptance [Heerink et al. (2009)], are emerging in HRI research.

As will be discussed in Sections 8.3 and 8.4, ethical experiment design and evaluation requires stakeholders to be identified, included, and addressed in the design and implementation of an HRI study. This includes creating study protocols and evaluation methods suitable for addressing the needs, values, and priorities of robot end-users and relevant communities, such as healthcare workers and patients and their families. On the scale of a single HRI study, quantitative and/or qualitative methods can be used to collect data representing the use and acceptance of robotic systems, for instance, adopting value-sensitive design methods [Friedman et al. (2017)] to investigate how healthcare robots can respect a patient's dignity [Van Wynsberghe (2020)]. Quantitative methods, including the measurement of objective task performance or the use of standardised questionnaires, can provide numerical evidence about the utility, perception, and relevance of the robot in the proposed scenario.

In addition to quantitative methods, qualitative evaluation and mixed-method approaches can be useful in HRI studies to provide contextualised and in-depth analysis. In Chapter 11, we will provide an overview of common qualitative methods for HRI studies.

Recommendations for Constructs and Metrics for HRI Evaluation

See Table 5.2 for a quick overview of the common constructs and measurements. An experiment may use a combination of multiple measurements to evaluate the same HRI factor. Convergent results based on multiple metrics that all support the hypothesis is the strongest evidence, as each measurement has inherent weaknesses and biases.

1. **Task performance metrics** should not to be used alone to substantiate claims about the entire HRI experience.

2. **Behavioural measurements** quantify human-related constructs. When using these measurements, psychology effects must be taken into account, with sufficient training given to the raters. Behavioural measurements can be used in conjunction with other metrics.

3. **Psychophysiological measurements** give detailed information about the participant's internal state. Significant expertise is required to ensure that the correct measurements are taken and the right interpretations and conclusions are drawn.

4. **Self-assessments data** are often collected through the use of surveys and questionnaires. Self-assessments can also be influenced by various human factors. Whenever possible, it is better to use existing questionnaires because they have been validated.

5. **The design of a questionnaire** involves selecting what to measure and how to measure it. Psychometric validation is lengthy, complex, and time-consuming. Researchers should carefully consider the need to create a new scale and investigate other literature for suitability. If the scale is used for the first time, report it as part of the measures section. Include reports on survey validity and reliability of the survey when reporting the scale in a research trial.

Other considerations to be reported when describing measurements:

1. **Metric reliability** captures the quality of the data with respect to measurement errors. Test-retest reliability, internal consistency reliability, inter-rater reliability, and intra-rater reliability are common measures that are applicable to HRI studies.

2. **Validity** is the extent to which the data-collection tool measures the intended measurement. face validity, content validity, criterion validity, and discriminant validity are four important metrics to consider when exploring metric validity in the HRI setting.

6

Quantitative Data Collection and Analysis

Once researchers have decided on the experimental design of their user study, their next step is to collect and analyse the data to test the research hypothesis. To ensure that the conclusions drawn from the analysis are valid and generalisable, an appropriate statistical analysis procedure for the study design and data must be chosen and the validity and generalisability of the results (e.g., extreme metric values due to outliers or insufficient statistical power) should be confirmed. This chapter focuses on evaluation and quantitative data analysis for HRI research.

Specifically, we provide an overview of the statistical methods most frequently found in the literature as well as some recommendations on how to best use these methods. We also cover general recommendations on how to validate the data-collection process before the main study is conducted via pilot studies and approaches to clean, better understand, and post-treat the data before any statistical analysis. The recommendations listed in this chapter are common practices in fields such as psychology [e.g., Leong and Austin (2006)] and human-computer interaction [e.g., Lazar et al. (2017), and Hornbæk (2013)]. Examples that showcase how statistical analyses are commonly done in HRI studies are also presented across the different sections of this chapter.

6.1 Pilot Studies

Pilot studies are small-scale versions of a full study, often conducted with one of two objectives: evaluate the feasibility of a major study or test whether the components and attributes (e.g., the pragmatics of participant recruitment, the suitability of a measurement instrument, and the validity and trustworthiness of the data that may be obtained using a particular tool) of the main study can all work together.

DOI: 10.1201/9781032693682-7

Although the small, exploratory nature of pilot studies makes them unsuitable for determining the validity of a hypothesis or the magnitude of an effect [Kraemer et al. (2006)], they still play a critical role in the planning of larger studies. Pilot studies provide researchers with an opportunity to identify modifications needed in the design of the study or system and provide insights into whether the study or system design conforms to the previously identified ethical and functional requirements. Besides helping researchers identify and correct potential deficiencies with their research design prior to the implementation of the full study, pilot studies can also help all the members of the research team familiarise themselves with the study protocol and assess the appropriateness of the planned data collection and analysis techniques [Van Teijlingen and Hundley (2002)]. In the case of HRI studies, pilot studies have helped identify potential problems with the data-collection methods [e.g., Bethel and Murphy (2010) and Bethel et al. (2007)] and the design of new measurement instruments [e.g., Rueben et al. (2020a)].

Pilot studies can also help identify potential problems with the data-collection methodology. For example, Bethel and Murphy (2010) were able to identify that participants often misunderstood questions relating to the dominance dimension when using the Self-Assessment Manikin (SAM) [Bradley and Lang (1994)] to report their affective state. Based on this observation, the authors later decided to exclude participants' dominance ratings from their analysis. Pilot studies also help to identify serious omissions or mistakes in data collection [Hornbæk (2013)]. This is of particular importance when psychophysiological measurements are included as part of a study since the quality and reliability of these measurements are often dependent on factors such as noise, lighting, and sensor placement [Bethel et al. (2007), Tiberio et al. (2013)]. Similarly, when a new questionnaire is developed or modifications are applied to an existing validated questionnaire, it is often recommended that the researchers pilot test the questionnaires. In this context, pilot studies will help researchers to test the duration of the questionnaire as well as identify ambiguous or difficult questions [Rueben et al. (2020a)].

In quantitative studies, pilots help determine whether the independent variables or conditions being manipulated by the researchers work as intended [e.g., Homburg (2018)] and assist in detecting confounds, that is, factors that can adversely affect the relationship between the independent and dependent factors such as the novelty effect, the physical embodiment of a robot, or the environment in which the main study will take place [Baxter et al. (2016), Feil-Seifer and Mataric (2009)]. For instance, if a study aims to investigate how robots can convey an affective state through their movement, a pilot study can confirm that participants can recognise the robot's gestures and postures [Homburg (2018)].

In qualitative studies, pilots help researchers preempt possible challenges in data collection and analysis (e.g., risks to the researchers themselves and participants) [Williams-McBean (2019)]. They increase researchers' confidence in the use of the different data-collection instruments as part of the main study. Pilot studies are valuable in uncovering ethical issues (e.g., a particular robot action that could harm the participants' safety during the main study) and potential researcher biases (e.g., instances where the researcher has a strong affinity for the participants taking part in a study) or expectancies. They can also help researchers identify barriers related to recruiting potential participants, particularly when working with vulnerable populations such as children or patients.

As pilot studies are smaller versions of the main study, they should closely follow the research design and protocol of the main study. For instance, the participants involved in the pilot study should be representative of the target study population and their selection should be based on the same inclusion and exclusion criteria as the main study [Thabane et al. (2010)]. Similarly, pilot studies should be executed using the same administration and measurement procedures that would be used to carry out the real study [Salkind (2010)]. A comprehensive pilot testing phase assesses all parts of the experiment until stability is achieved and any critical systematic bias (e.g., learning effects and novelty effects) has plateaued. Thus, several rounds of pilot testing might be required. Finally, it is important to note that the success of the main study is not guaranteed by having one or more pilot studies. It is still possible that problems not previously encountered during the pilot testing phase might arise during the actual study [Van Teijlingen and Hundley (2002)].

Recommendation: In addition to the careful design and planning of a research study, it is beneficial for researchers to conduct a good pilot study prior to the main study. This will help them identify critical problems and deficiencies in the study protocol as well as plan and take necessary corrective actions.

6.2 Considerations Prior to Data Analysis

No matter how well a user study is designed and implemented, researchers frequently have to deal with errors and their effects on study results. In this subsection, we describe the strategies for data cleaning and outlier detection, appropriate visualisation techniques to better understand and examine data, and the selection of appropriate statistical analyses.

6.2.1 Data Cleaning

Data cleaning can be described as an iterative and repeated three-stage process in which data errors and abnormalities are screened, diagnosed, and edited. In most studies, data cleaning is mainly done as a post data-collection process. However, careful data monitoring and visualisation during the collection process can also help to improve the quality of the data. It is not always immediately clear whether a suspected data point is erroneous and why it is so. Similarly, missing values could be due to interruptions during data collection or unavailable information [Van den Broeck et al. (2005)]. Thus, researchers may utilise their knowledge about potential technical errors (e.g., measurement errors) or previously identified systematic biases (e.g., novelty or learning effects) and expected ranges of normal values to define the rules to follow during data cleaning.

6.2.1.1 Screening

A good starting point for detecting invalid data is to do a visual scan of each participant's responses and measurements as they are collected. Patterned responses in questionnaires or responses completed in less time than what was observed across all participants are often employed as indicators of invalid data [Leong and Austin (2006)]. If instructional manipulation checks or attention check questions, that is, questions with obvious answers, were included among the measurements collected during the study, they can also be employed to identify potentially erroneous data points. This screening method is frequently used in studies where participants were recruited through crowd-sourcing services [Hauser and Schwarz (2016)]. During screening, researchers can also look for lack or excess of data (e.g., missing values), outliers, and strange patterns in the data distributions.

Most methods used to evaluate possible outliers are based on the idea that a particular proportion of valid data points will exist within a given k number of standard deviations from the population mean. By setting a threshold on the number of standard deviations considered valid, data points above and below that threshold are considered to be outliers. Statistics such as the first (Q1) and third (Q3) quartiles of the data as well as the interquartile range (IQR = Q3 − Q1) are also used to define outlier detection bounds. Data points outside these bounds are considered to be outliers and are thus eligible to be excluded from any posterior data analysis.

In the case of a sample size that is relatively small, alternative methods such as Grubb's test [see Leong and Austin (2006) for more information] are recommended instead. When looking for outliers, it is important to keep in mind that although all outliers are characterised as extreme data points, not all extreme data points fall into the outlier category. Similarly, before deciding to exclude potential outliers from any future statistical tests, it is important to determine whether these data points are distributed equally across conditions. Otherwise, we run the risk of removing valid observations and experimental effects.

6.2.1.2 Diagnosis

In this phase, researchers clarify the nature of potentially problematic data points identified during data screening. Data points can be categorised as erroneous, true extreme, true normal (i.e., prior expectations about normal data ranges were incorrect), or idiopathic (i.e., a data point that is suspected to be in error but lacks an explanation for why it might be incorrect). For the suspected points for which diagnosis is less straightforward, that is, they do not fall into the expected ranges of true extreme or outlier values, the application of a combination of diagnostic procedures is recommended [Van den Broeck et al. (2005)]. These diagnostic procedures could be to determine whether data points were consistently the same throughout the whole data-collection procedure or to collect additional information, for example, question the experimenter in charge of data collection, and if possible, repeat the measurement.

6.2.1.3 Editing

After the identification of errors, missing values, and true values, the next step is to decide what to do with these problematic data points. Problematic data points can be corrected, excluded, or left unchanged. In the case of impossible values, the general recommendation is to correct them if a correct value can be found or exclude them otherwise [Leong and Austin (2006)]. This editing rule also applies when missing values are due to hardware or software failures. In the case of missing data, the researcher must determine the acceptable amount of missing data before leaving a participant's data out of the analysis, as well as what acceptable substitute values to use instead (e.g., sample-wide median values are often employed to substitute a missing variable or metric) [Leong and Austin (2006)]. Additionally, researchers need to declare excluded data and the reasons for excluding this data.

Whatever the data-cleaning strategy researchers employ prior to data analysis, they should provide detailed documentation of the data-cleaning methods, error types and rates, and error deletion and correction rates in their study report. Similarly, it is necessary to report on the differences in the outcomes of their studies with and without outliers [Van den Broeck et al. (2005)].

6.2.2 Data Visualisation

Simple visualisations can be produced for both individual participants and the aggregated responses, with the goal of seeing the trend (the direction of data progression over time), level (changes in the relative value of the data over the dependent variable), and stability and variability of the data between experimental conditions and participants [Lane and Gast (2014)]. Four common techniques are highlighted in this section: cumulative records, semi-logarithmic charts, bar graphs, and line graphs.

Cumulative records are generated by summing the participants' responses across sessions. This method is useful when there is a progression in the experimental conditions (e.g., in each experiment, the robot learns to do a new skill in addition to what it could do previously) and a survey is administered at the end of each condition. This type of graph is a simple way to illustrate trends (increasing, decreasing, and no change) in the total response within a study.

Semi-logarithmic charts have one axis on a linear scale while the other on a logarithmic scale. They are commonly used to display the rate of change or proportional changes in performance. These graphs are common in machine learning to track the accuracy of the algorithm over the training iterations. Both cumulative records and semi-logarithmic charts are used for continuous data, whereas bar graphs are commonly used for discrete data or continuous data for studies with small sample sizes [Lane and Gast (2014), Weissgerber et al. (2017)].

Bar graphs are typically used for presenting summary statistics between experimental conditions (e.g., user ratings between the control condition versus the robot condition). However, summary statistics can be problematic because many distributions can lead to the same graph while other important features of the dataset are hidden [Weissgerber et al. (2017), George et al. (2017)]. In the case of a small sample size, it is recommended that the researcher plot the data distributions or use scatter plots instead of traditional bar graphs [Weissgerber et al. (2015)]. Weissgerber et al. (2017) implemented a free, online tool for data visualisation (http://statistika.mfub.bg.ac.rs/interactive-dotplot/), which can serve as a starting point for researchers when deciding between different visualisations.

Line graphs are often used when there is a temporal aspect to the data (e.g., motion tracking during an HRI experiment) [Lane and Gast (2014)].

6.3 Data Analysis

When analysing quantitative data, researchers generally have one of two objectives:

1. Determine the validity of a hypothesis by comparing population estimates between groups (e.g., whether the acceptance ratings of a robot change between female and male participants) or effects

2. Analyse the relationship between a dependent variable (e.g., a task performance metrics or the final score of a subjective questionnaire) and independent variables, or *predictors* (e.g., two different user interfaces for robot control and teleoperation).

Hence, quantitative analysis methods can also assist HRI researchers in the validation of the functional requirements of the HRI systems they design.

This section provides an overview of the most common statistical analyses found in the literature: hypothesis testing methods and confidence intervals. We also briefly discuss what statistical analyses are recommended when dealing with ordinal data, such as the one obtained when using Likert scales.

6.3.1 Hypothesis Testing

Hypothesis testing is the most common approach used in HRI studies. It consists of a formal approach for deciding between two interpretations of a statistical relationship in a sample given a chosen threshold. One interpretation is called the *null hypothesis* (often symbolised as H_0), which refers to no relationship in the target population. The other interpretation is called the *alternative hypothesis* (often symbolised as H_1), which posits that there is the existence of a significant relationship between variables in the target population that is reflected in the population sample as well. The significance level is the threshold at which it is decided whether to reject or retain the null hypothesis. This significance value is also known as the *p-value* and is related to the probability statement made about the observed sample in the context of a hypothesis, not about the hypotheses being tested [Altman and Krzywinski (2017)]. The smallest significance level that is normally considered as reasonable evidence is 5%. In practice, if the probability of observing the *null hypothesis* statement H_0 is less than 5%, the *null hypothesis* is rejected in favour of the *alternative hypothesis* H_1. The *p-value* represents the probability of observing a similar statistical relationship if the data being analysed was generated from random samples [Kaptein and Robertson (2012)].

Hypothesis testing methods can be classified into parametric and non-parametric tests. These two classes are discussed in detail in the following sections. We will use four exemplar HRI studies to illustrate hypothesis testing approaches. In the first example study, Tanevska et al. (2020) compared human perceptions of the iCub robot when it uses personalised versus adaptive behaviour in three different settings. They hypothesised that the adaptive behaviour would enrich the HRI compared to the non-adaptive one. In the second example study, Sena and Howard (2020) proposed a visual feedback framework in the robot learning from demonstration that allows the novice user to observe robot learning progress and provide better distribution for the demonstrations over the task space. They hypothesised that visualisation of the robot's learning progress would result in a significant improvement in teaching efficiency, compared with a no-guidance teaching process. In the third example, Szafir (2019) compared different interface designs using Augmented Reality (AR) in a human-drone interaction study. Szafir represents the drone's motion intention using four paradigms and compared them with the baseline where there is no augmented information at all. He hypothesised that using AR can address communication problems such as incorrect mental models, poorly calibrated trust, and inadequate situational awareness. In the fourth example study, Akalin et al. (2019) examined different feedback styles from a social robot to older people in robot-assisted training. They compared the effect of flattering, positive, and negative feedback on robot acceptance and task performance. They hypothesised that both the flattering and positive feedback will be appreciated by the users, leading to improved task performance.

6.3.1.1 Standard Parametric Methods

Parametric tests are widely used in analysing HRI data. These tests include t-tests, analysis of variances (ANOVAs), and ordinary least squares regressions. Here, we focus on the ANOVA test because it is the most well-known and used test. It is also the most misused test. The ANOVA test assesses the potential difference between two or more groups on a *continuous* measurement. Before using ANOVA, some assumptions need to be met in the tested data [Cardinal and Aitken (2013)]. These assumptions are as follows:

1. Normality of data samples
2. Homogeneity of variance (sphericity)
3. Independence of samples

There are different tests for normality, such as chi-square, Kolmogorov-Smirnov, Shapiro-Wilk, Jarque-Bera, and D'Agostino-Pearson. Some researchers argue that if the sample size of all groups are equal (a balanced model) and sufficiently large, the normality assumption can be relaxed provided the samples are symmetrical [Blanca et al. (2017), Feir-Walsh and Toothaker (1974)]. The homogeneity assumption can be tested by comparing variance values or through specific statistical tests, such as Levene's, Fligner Killeen, and Bartlett's tests. ANOVA is not robust against unequal variances [Grissom (2000)], thus if this assumption is violated, it may alter both the type I error, that is, detecting an effect when there is none [Harwell et al. (1992)] and the statistical power of the test [Nimon (2012)]. For the independence assumption, there is no specific test that proves its validity, but it needs to be accounted for in the design of the study. For instance, the observation data may be dependent if repeated measurements are collected from the same subject.

From the study design perspective (Section 3.1), ANOVA can be categorised based on the study model as well as the number of independent variables considered in the study [Tabachnick et al. (2007)]. In the following, we list the different types of ANOVA:

1. *Between-subjects ANOVA* is used when examining the differences between two or more independent groups. Within this type of ANOVA, there are one-way ANOVA and factorial ANOVA. The main difference between these two branches is the number of independent variables to be tested.

   (a) *One-way ANOVA* is used when assessing the difference between groups with respect to one independent variable. In practice, one-way ANOVA is used when studying the difference between at least three groups because the t-test is usually used when there are two groups. Szafir (2019) used one-way ANOVA to compare five groups (i.e., baseline and four different interface designs) with respect to one variable: task performance.

   (b) *Factorial ANOVA* is used when examining multiple independent variables.

2. *Within-subjects ANOVA*, also called repeated measures ANOVA, is used when the same subjects were tested for each experimental condition. It is frequently used with pre- and post-test experimental design. However, it is not limited to only two time periods; it can also be used when examining the differences over two or more measurements in time. When using this type of ANOVA, it is also important to test the sphericity assumption, that is, the assumption that the variance in the differences between all pairs of groups are equal is valid [Mauchly (1940)]. This assumption can be tested using Mauchly's test. Sena and Howard (2020) checked and verified that their collected data met the following assumptions before conducting a repeated measures ANOVA test: the data was normally distributed and continuous, there were no outliers, and the dependent variables fulfilled the assumption of sphericity. The repeated measures ANOVA was used to compare among three different visual feedback methodologies that all users were exposed to (within-subjects).

3. *Mixed-model ANOVA*, also called within-between ANOVA, is used when studying the differences by group and time. That is, when the study follows both a between-subjects and within-subjects design. Tanevska et al. (2020) used a mixed model to test their hypothesis by comparing within subjects in different sessions as well as comparing between subjects interacting with the iCub robot in different interaction settings.

4. *Multivariate analysis of variance (MANOVA)* is used when studying the differences between groups on multiple dependent variables. We can not replace it by simply performing multiple ANOVA for each dependent variable because ANOVA would not account for the correlations between the dependent variables.

6.3.1.2 Post-Hoc Tests and Corrections

While the ANOVA test determines whether the differences observed among groups are due to chance, it does not identify which particular differences between pairs of groups are significant or not significant. This is what *post-hoc* tests do; they are used after the ANOVA test to assess the differences between all possible group pairs (multiple comparisons) [Hsu (1996)]. The more groups in an ANOVA test, the higher the type I error rate. The type I error rate (i.e., false positive) is defined by the significance level in the case of one comparison (two groups). The error rate inflates with the increase in the number of groups since it increases the number of comparisons [Thompson (1994)]. A post-hoc test constrains the experiment-wise error rate to the significance level with an adjusted p-value. A variety of methods exist in the literature for conducting post-hoc tests. Here, we focus on the most common tests available in statistical software.

- *The Bonferroni test* is simply a number of t-tests performed on all possible pairs of the tested groups. To limit the type I error rate to the significance level, the Bonferroni test sets the significance cutoff at the significance level divided by the number of comparisons. This is called *Bonferroni correction*. Thus, the Bonferroni test tends to be more conservative. Tanevska et al. (2020) utilised the Bonferroni test after significance was found in mixed-model ANOVA analysis.

- *Tukey's Honest Significant Difference (HSD)* test is used to measure the pairwise differences in groups. It gives an estimate of the difference between the groups and a confidence interval for the estimate. This test is used with balanced data, that is, when the sample size is equal between groups. Akalin et al. (2019) employed the Tukey HSD test, after significance was found in one-way ANOVA analysis for multiple comparisons between different feedback styles and their effect on task performance.

- The *Tukey-Kramer HSD test* is the same as the Tukey HSD test except that it can be used with unbalanced groups.

- *Fisher's Least Significant Difference (LSD) test* is the most "liberal" method because it has a high probability of type I error rate. This test is computationally identical to the Bonferroni test except it does not employ any adjustment in type I error [Sheskin (2020)]. Bethel (2009) employed Fisher's LSD test after significance was found in a two-way ANOVA analysis for comparing people's perception of a robot's different operating modes.

- The *Student-Newman-Keuls* (SNK) *test* orders the differences in the "group" means from the largest to the smallest in a step-wise procedure. Then, it starts with the largest difference to test if it is significant; if so, the test continues in order until it reaches a non-significant mean difference, at which time the test is terminated. It is more liable to type II error (i.e., false negative) than a type I error [Hilton and Armstrong (2006)].

- The *Duncan's Multiple Range test* is similar to the SNK test, but the probability of making a type I error increases with an increase in the number of comparisons.
- *Scheffe's test* is a very conservative method that gives the highest protection against type I errors. Also, it is robust to the violations of the ANOVA's assumption [Armstrong and Hilton (2004)]. However, all of these come at the cost of less power for detecting effects.
- *Dunnett's test* is used when we care about the difference between the control group and the other groups; not considering all the pairwise comparisons. Also, it can be used with unbalanced groups as well as if the homogeneity assumption is violated. Szafir (2019) employed Dunnett's test after significance was found in their one-way ANOVA analysis.
- The *Binomial Sign test* uses only the signs of the differences between the pairwise comparisons.

For more details about post-hoc tests and types, we refer the reader to Homack (2001), Hilton and Armstrong (2006), and Cardinal and Aitken (2013).

6.3.2 Confidence Interval

A confidence interval (CI) uses the same underlying mathematical methods as hypothesis testing, but instead of giving a probability of a single value, a range of values is indicated [Dix (2020)]. It is powerful for showcasing the level of uncertainty around an estimate or prediction. The confidence interval is defined as a range of values, calculated by statistical methods, that includes the desired true parameter with a probability, called the *confidence level*, defined in advance. The size of the confidence interval depends on the sample size, the standard deviation, and the selected level of confidence [Gardner and Altman (1986)]. For instance, if the sample size is large, the confidence interval will be narrow. On the other hand, the larger the confidence level is, the wider the confidence interval. The most commonly used value for the confidence level is 95% – similar to a 5% level of statistical significance in the hypothesis test.

CIs can be one- or two-sided. A two-sided CI defines the population parameter from both lower and upper bounds. A one-sided CI provides either an upper or a lower limit to the population parameter. Calculation of the CI of a sample parameter takes the general form of CI = Point estimate

± Margin of error where the margin of error is given by the product of the critical value, selected as per the required confidence limits and the standard error of point estimate. The CI is reported with the point estimate of the concerned parameter, indicating the reliability of the estimate. Hancock et al. (2020) utilised the CI for analysing the effects of human, robot, and environmental factors on the perceived trust in HRI. They used a two-sided CI for each factor, and they found that most of the factor comparisons did not include zero. This suggested that the identified relationship is consistent and substantive.

6.3.3 Statistical Analysis Based on the Type of Measurement Scales

Most of the quantitative measurements obtained in HRI studies can be categorised into three types: continuous, interval values (e.g., behavioral measurements such as the distance between a human and a robot or physiological data such as heart rate), and ordinal data (e.g., Likert items and scales often used on self-reporting questionnaires). While the choice of statistical methods to analyse continuous and interval values is a more or less straightforward process[1], there is an ongoing debate on what is the correct way of analysing ordinal data.

On the one hand, some researchers argue that ordinal data obtained using Likert scales can be treated as interval data and thus standard parametric statistical methods can be used. On the other hand, more conservative researchers choose to employ non-parametric tests instead even though it is known that these methods are less powerful and lack sensitivity to detect smaller effects [Sullivan and Artino Jr (2013)].

Recently, Schrum et al. (2020) provided a set of recommendations on the use and analysis of Likert scales in HRI studies:

1. Employ summary statistics such as mode, median, range, and skewness when reporting individual Likert items.

2. Although there is compelling evidence suggesting that parametric tests such as ANOVA can be used on Likert scale data [Sullivan and Artino Jr (2013)], as long as the appropriate assumptions have been tested and validated, a conservative approach in which non-parametric tests are used instead is recommended.

3. Perform analysis on a multi-item scale instead of on single Likert items.

Recommendation: Figure 6.1 provides a decision flow chart to aid researchers to choose suitable statistical methods for their HRI study. However, researchers should not blindly apply statistical hypotheses tests, such as ANOVA, to seek statistical significance. One trap researchers often fall into while conducting quantitative experiments is that of being obsessed

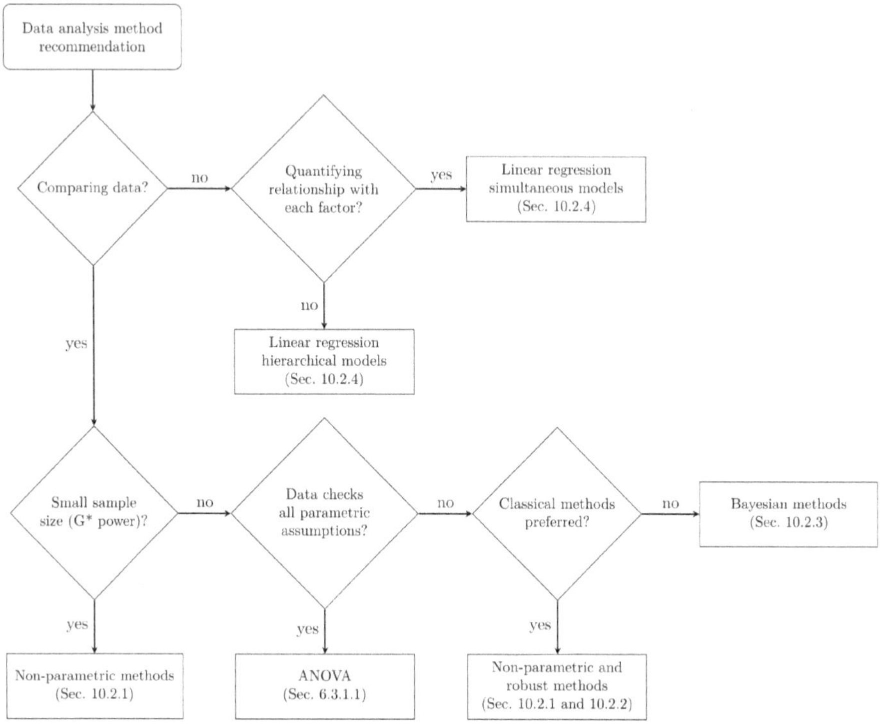

FIGURE 6.1
A decision flow chart for choosing the statistical method that best fits the characteristics of the data to be analysed.

with the p-value, which is used to reject the null hypothesis. Researchers can potentially fall into the trap of determining the success or failure of our studies by whether the p-values given by an ANOVA analysis are under the significance criterion of $\alpha = 0.05$. If a study is well-designed, even if the final p-value is not significant, it conveys meaningful results that suggest there is no strong evidence of the causal relation being investigated. For a well-designed experiment, researchers pay more attention to statistical power analysis that includes the relationship between the sample size and effect size. Proper experiments ensure that the power will be reasonably high to detect the significance with sample size calculation [Kim et al. (2014)]. Despite the potential misuse, there is a better general understanding of classic parametric methods. Researchers are more likely to use them correctly, and readers are more likely to understand the results better. It is important to report p-values as well as confidence intervals and effect sizes when possible [Dix (2020)]. Finally, it is important to include all the data necessary to reproduce your results (e.g., means, variances, number of participants, demographics, number of conditions, and excluded data points) [Bartneck et al. (2020b)].

6.4 Examples

Sena and Howard (2020) proposed a visual feedback framework in robot learning from demonstration that allows the novice user to observe the robot learning progress and provides better distribution for the demonstrations over the task space. They compared among three conditions for the visual feedback methodology using a within-subjects study. They conducted a power analysis prior to the data collection that indicated that for a medium effect size (Cohen's $f = 0.3$) and a type I error rate $= 0.05$, the required sample size is 30. Then, prior to performing a repeated measures ANOVA analysis, they checked the collected data to verify the normally distributed dependent variable, the continuous dependent variable, the absence of outliers, and the sphericity of the data (homogeneity of variance). They also conducted another study that compared among four conditions, and they used mixed-factor ANOVA analysis. The power analysis indicated the required sample size was 36, or 9 per condition. However, when they checked the ANOVA assumptions, they found that the data for three out of four conditions were not normally distributed using the Anderson-Darling test. The excess kurtosis for the four groups was 1.37, 1.60, –0.64, and 0.98, where a normal distribution would have an excess kurtosis value of 0. They chose to go with the argument about the robustness of the ANOVA for the violation of the normality assumption especially with a *minor* violation. They also tested for the homogeneity of the variance using Mauchly's test and found $\chi^2(2) = 1.24, p = 0.54$, which indicate the validity of this assumption and that data correction is not needed.

6.5 Summary

In this chapter we provided an overview of collecting and analysing quantitative data. We also covered common hypothesis testing approaches. However, this discussion here only covered a subset of HRI data analysis methods. In Chapter 10, we will further discuss advanced data analysis approaches, such as Bayesian statistics and small sample size studies. We will also discuss potential biases in data. In addition, HRI studies can also generate qualitative data. In Chapter 12, we will discuss the analysis of qualitative data.

Quantitative Data Collection and Analysis of HRI Studies

1. **Piloting**: Conduct a pilot study that covers the proposed scope of the experiment prior to the main study. This will help researchers identify critical problems and deficiencies in the study protocol as well as plan and take all the corrective actions.

2. **Quality of data**: Data cleaning and post-processing is an important step prior to data analysis since the validity and generalisability of the observations and conclusions made in a study strongly depend on the quality of the data being analysed. Formulate a set of pre-defined rules for dealing with data errors and missing or extreme values beforehand. It is also important to provide detailed documentation of all the data-cleaning methods employed, the types and sources of error identified in the data, as well as the excluded data and the reasons for exclusion. We note also that data can vary widely on how much pre- and post-processing they require prior to data analysis. For instance, signal data obtained from physiological measurements, such as gaze, require more filtering and pre- and post-processing than do questionnaire answers.

3. **Reproducibility**: It is important to report all the details necessary to reproduce the results of statistical analysis (e.g., means, variances, number of participants, demographics, number of conditions, and excluded data points).

Statistical Analysis of HRI Study Data

- Look at the data before jumping to the statistical testing stage. Visualisation techniques, such as scatter plots, allow us to get an initial idea about the spread of the data, anomalies or extreme values, and the match between the data and the model (i.e., is the data linear?)
- Avoid blindly applying statistical hypothesis tests, such as ANOVA, seeking for statistical significance. If a study is well-designed, even if the final p-value is not significant, it conveys meaningful results that suggest there is no strong evidence of the causal relation being investigated. A well-designed experiment requires attention to statistical power analysis that includes the relationship between the sample size and effect size.
- Proper experiments ensure that the power will be reasonably high to detect the significance with sample size calculation. Also, it is important to report p-values as well as confidence intervals and effect sizes when possible.
- Despite the potential misuse, there is a better general understanding of classic parametric methods. Researchers are more likely to use them correctly, and readers are more likely to understand the results better.
- To help to decide on what statistical method to apply when, Figure 6.1 shows a decision map that guides the reader to the most suitable test given the purpose of their analysis and the nature of the data to be analysed.

Note

1. The literature suggests that classic parametric tests (e.g., ANOVA) are preferred if assumptions of normality and sphericity are valid, otherwise non-parametric tests (e.g., Kruskal-Wallis) are used instead [Venero and Davila (2020)].

7

Reporting Findings and Designing Follow-up Studies

This chapter discusses the post-study phase of HRI research at the end of a study life cycle, including considerations regarding reporting, the iterative design process, the development of follow-up studies, and approaches to support open-sourcing, bench-marking, and replication efforts.

7.1 Transparent and Reproducible HRI

Increasing the reproducibility of HRI studies benefits continuous progression of the field by allowing new research to directly build on existing efforts [Bonsignorio and Del Pobil (2015), Gunes et al. (2022)]. Ullman et al. (2021) discussed two types of HRI replication studies, namely, direct replication of an original study, and adding one variable (e.g., a different robot) and conducting a between-subjects study to compare with the original study. Clear and transparent reporting of an HRI study is required to aid such direct replication or extension as a between-subjects study.

7.1.1 Recommendations on Transparent Reporting

When reporting on the methodology of an HRI study, the goal is to provide a clear description with sufficient details to allow replication of the experiment. This may include the study context, participant demographics, evaluation approaches and design variables, and data collection, cleaning, and analysis methods. Bagchi et al. (2023) provide recommendations to formalise reporting in HRI studies to support meta-analysis and reproducible research. They emphasise the requirement to provide detailed information on participants and recruitment to reveal potential biases and encourage ethical research practices, including target and actual demographics, how demographics information was collected, and inclusion and exclusion criteria. Further, they suggest that structured descriptions of key elements in the

DOI: 10.1201/9781032693682-8

study design and execution and release of code and data can increase the reporting transparency. An online template (https://tinyurl.com/hri-study-reporting) accompanying the recommendations by Bagchi et al. (2023) is available for researchers to employ when reporting on their study design.

As publication venues often have a page or word limit, researchers may choose to include supplementary materials for detailed description or a copy of study materials. Some examples may be a copy of the questionnaires or video demonstrations of a robot's interactive behaviours. For example, Tijani et al. (2024) extended the Choreographe graphical programming interface to support generalised language-based HRI experiments using the Pepper and NAO robots. Demonstration of an example behaviour implemented with the proposed tool on a Pepper robot was provided as a footnote on the YouTube video link. The proposed tool was open-sourced on Github with a link to the repository provided in the paper's abstract.

An especially encouraging trend in reporting HRI studies is open-sourcing of code and data, which enables bench-marking efforts. This is similar to practices in the machine learning community, for example, [Papers with Code (2024)], that increases reproducibility and objective evaluation of the state of the art. However, as HRI studies are often conducted with a specific robot platform and programming API, open-sourced code and models may not be directly applicable to a new study using a different platform. Gunes et al. (2022) discussed such technical challenges in supporting HRI reproducibility in their review, including availability of the robot platform and software artefacts. Open-sourcing the code may also incur additional efforts from the researchers to continuously maintain the repository to support adoption and replication by other researchers. Further, HRI data collected via user studies may contain identifiable or sensitive human data. We further discussed privacy considerations in HRI data collection and publication in Section 4.6.1 and will discuss them further in Section 8.3.3.

When reporting the results of an HRI study, it is important to position the observations in the specific context of the study design (see Section 3.2) to accurately interpret the implications and limitations of the findings. For example, previous studies have shown that perceptions towards a robot can vary across cultures [Lee and Sabanović (2014)] and age groups [Lee et al. (2016)]. By providing sufficient context, researchers can reduce the risks of over-generalising the conclusions. In an example replication study, Strait et al. (2020) investigated the joint Simon effect, that is, people working more effectively with a human-like robot compared to a mechanomorphic robot. Experiments were conducted at three sites with participants from different cultures, namely Germany, the U.S. Midwest, and the Mexico-U.S. border. The joint Simon effect was observed in these cross-site experiments with consistent implementation and experimental protocols. Thus, this study implies that the joint Simon effect may generalise across some cultures. In another example, Ullman et al. (2021) investigated human-robot trust with two in-person studies on two robot platforms ($n = 140$ and $n = 200$), as well as an online video comparison study with a large sample size ($n = 400$). Their analyses revealed that findings in an underpowered study may not generalise when the sample size increased. However, the additional cost and effort required to conduct multisite and large sample size studies may limit their applicability for HRI research.

In confirmatory studies (see Section 2.1.2), where a hypothesis regarding a specific measurement is tested with a statistical significance test (see Section 6.3.1 on null hypothesis testing), one consideration is to avoid p-hacking, that is, a researcher bias in which data were collected or selected until non-significant results become significant in the statistical tests. This is especially important in small sample size studies, which are common in HRI research due to resource constraints. In fields such as biology and psychology, there have been surveys and guidelines on avoiding p-hacking [Head et al. (2015), Wicherts et al. (2016)], which HRI studies may also benefit from. In Chapter 10, we will further discuss common biases and errors (Section 10.1), advanced statistical methods (Section 10.2), and small sample size studies (Section 10.3).

7.1.2 Creating a Supportive Culture for Reproducible HRI

Gunes et al. (2022) reviewed practices and challenges of study reproducibility in HRI research based on the HRI 2020 conference with a theme in reproducibility. They identified biases in the review culture and research practices that created difficulties for reproducible HRI studies. Specifically, normative statistical bias in Null Hypothesis Significance Testing (NHST) can discourage exploratory and qualitative HRI studies, as well as HRI studies with a small sample size. In Chapter 10, we will discuss statistical analysis alternatives to NHST and considerations on small sample sized HRI studies. Further, Gunes et al. (2022) reported that reviewers can exhibit a range of biases when evaluating a paper submission, including: novelty bias, that is favouring studies investigating a new concept instead of replication studies), positive results bias, that is favouring studies that report results confirming the hypotheses), and physical robot bias, that is favouring experiments with physically embodied robots). In addition, a number of societal systemic biases were reported to influence HRI research practices. Specifically, the emphasis on physical robots can prevent researchers without access to the hardware (i.e., classism bias) or participants from certain groups (i.e., racial, ethnic, and cultural sampling biases) from contributing to and participating in HRI research. Further, there are robot morphology biases that may reinforce systemic societal biases. In Section 10.1 we will further discuss common sources of bias in HRI research.

To create a supportive culture for reproducible HRI studies, Gunes et al. (2022) suggested enhancing the training of HRI researchers to embrace "failures" and results that may not confirm the hypotheses, to acknowledge reproducibility in quantitative and qualitative studies, to practice pre-registration in study design and planning, and to provide revised guidance for authors, reviewers, and editors to reform the publication and review process of HRI research. As was discussed in Section 2.4, pre-registration of research questions and hypotheses can increase the reproducibility of an HRI study. By pre-registering the study plan, researchers can reduce biases introduced by post-hoc evaluation of outcomes. A growing number of journals, such as the *International Journal of Social Robotics*, and conferences are also requiring publications to include ethics statements, data availability statements, clarifications on funding sources, and conflict of interest statements. This increases transparency about the potential benefits and risks of a study, as will be further discussed in Section 8.6.

7.1.3 Tools for Reproducible HRI

A number of tools and frameworks have been developed to support reproducible HRI studies. For example, Groß et al. (2023) identified five key requirements for a software architecture to support the development of social dialogue behaviours in an interdisciplinary research team. This includes configurable interaction behaviours, a user interface explaining robot behaviours, autonomous and controlled robot action execution, extensibility for adding additional functions and sensors, and being agnostic to robot platforms or interaction modalities. Guided by these requirements, Groß et al. (2023) developed the open-sourced Robotics Integration and Scenario-Management Extensible-Architecture (RISE), utilising existing tools, such as the ROS middleware [Estefo et al. (2019)] and the Behavior Markup Language [Kopp et al. (2006)]. The proposed RISE system was tested with multiple robot platforms and interaction scenarios. The RISE software tool was released online under an MIT license with its documentation published online. In another example, Esterwood and Robert Jr (2023) presented an immersive simulator environment developed using the Unreal Engine. This tool was then used in experiments with two robot platforms to investigate human-robot trust repairs in warehouse robots. There have also been efforts in indexing available resources to support easy identification of HRI tools and datasets [see, for example, Chung (2021) and Shi (2024)].

While these tools were developed to support replication studies in HRI, they are not without limitations. There is often a high cost in developing and maintaining such software. Open-source tools developed as a research project can have reduced support and usability over time after the project has concluded or when the circumstances of the lead creators have changed. Commercial tools may maintain continuous support and improvement with a designated team of developers, while their licensing costs can create barriers for usage. Open-source tools developed by an active community have the benefits of both support and accessibility, as exemplified by the Robot Operating System 2 (ROS2) library [Macenski et al. (2022, 2023)], where the developers are often also users at the same time. Further, as HRI studies often involve complex experimental design with various hardware and software tools, the proliferation of different tools can hamper reproducibility.

7.2 Iterative Design and Follow-up Studies

The majority of HRI research studies are considered proof-of-concepts tested in the lab environment [Wang et al. (2019)], corresponding to low levels (1–3) on the NASA technology readiness levels (TRL) scale [Mankins et al. (1995)]. Low TRL systems are expected for an emerging research topic such as HRI, where many HRI works focus on fundamental research rather than application research, which is essential for scientific discovery. At the same time, efforts to increase ecological validity of study contexts is key to ensure robust and generalisable research outcomes. Adopting an iterative approach to build on previous work allows HRI researchers to create a scaffolding that incrementally expands their discovery or progresses application outcomes.

7.2.1 Increasing Ecological Validity in Follow-up Studies

As was discussed in Section 7.1.2, HRI replication studies following a confirmatory study can test generalisability of findings of the original study in extended contexts, such as with participants from different cultural backgrounds or with a different robot platform. In exploratory HRI studies, the findings often lead to new research questions and directions from which follow-up studies may spawn [e.g., Rueben et al. (2021)].

In addition, replication studies may focus on up-scaling an original study with increased sample size or deployment duration. Another approach to increase the ecological validity in a follow-up study is to replicate a lab-based experiment in a more natural setting with in-the-wild evaluation. We previously discussed study contexts in Section 3.2 and will provide further discussion in Section 9.2.3. Depending on the study aims and accessibility to resources, researchers are encouraged to incrementally increase the ecological validity of their study design in a series of HRI studies.

7.2.2 Iterative Design in HRI

Various design methods and tools have been adopted in HRI [Ostrowski et al. (2020), Lupetti et al. (2021)]. The iterative design process in particular can guide HRI researchers to develop a series of studies investigating a topic with growing depth or breadth. As widely adopted in the Human-Computer Interaction (HCI) literature [Kelley (1984)], the iterative design process captures a series of methodologies for designing and evaluating a new interaction technology. Kelley (1984) identified six empirical steps in the iterative design process, namely, task analysis discovering the requirements of users, deep structure development focusing on system implementation, a first simulation experiment using the WoZ approach (see Section 1.3.6), a first prototype of the program with approximation of the functions, a second WoZ experiment where the wizard provides interventions to the prototype program, and a cross-validation where the improved prototype is evaluated without wizard intervention using additional participants. Multiple implementation-evaluation-redesign cycles may be required to develop an HRI system or investigate an HRI research question.

In one example, Zamfirescu-Pereira et al. (2021b) discussed the benefits of exploratory prototyping in HRI with case studies adopting three methods. These included physical prototyping with human-in-the-loop control of robot behaviours, video prototyping using pre-recorded HRI encounters, and virtual simulations using virtual reality interactions. Comparing these exploratory prototyping methods, Zamfirescu-Pereira et al. (2021b) discussed the trade-offs between realism and cost, the advantage of immersion in naturalist environments and in-situ studies, and technical and engineering efforts required for HRI prototyping. Based on these case studies, they recommended HRI researchers adopt multiple methods to draw on their strengths. In another example, Knight et al. (2024) reported on an iterative design and evaluation of a robot waiter's delivery algorithms in a simulated cocktail event serving pairs of people. In the conceptualisation and design phase, the authors took inspiration from theatre techniques, with one participant acting as the robot waiter and two other participants acting as the patrons. This allowed an initial design of the research variables in robot behaviours and study contexts. Following this design, Knight et al. (2024) modified the hardware of an existing robot platform and implemented software required to realise the robot behaviour designs, which were then used in the HRI experiments.

While iterative design is an effective method to investigate a set of related HRI research questions, several challenges exist in designing and conducting iterative HRI studies. One major difficulty is the continuous involvement of stakeholders throughout a series of studies. In Section 8.2, we will discuss how to identify and engage stakeholders in HRI research. Further, in Section 11.1.7, we will discuss participatory design and co-design methods, in which stakeholders contribute to multiple phases of iterative

HRI research studies. Another challenge is to control for various experimental factors between the iterations, especially in follow-up studies designed to replicate most experimental protocols and compare with previous work as a between-subjects study, as was discussed in Section 7.1. In addition, limitations and biases in previous work may propagate through the iterations. Thus, researchers are encouraged to critically review the design and findings in a study when developing follow-up works based on it.

7.3 Summary

We discussed recommendations on reporting HRI study results in a transparent and reproducible manner. Further, adopting iterative design can increase the ecological validity of HRI research via a series of follow-up studies.

Post-Study Phase: Reproducibility and Follow-Up Studies

- Researchers are encouraged to clearly report their study methodology to enable others to replicate the experiments. Transparent reporting of study context, participant demographics, and study execution is important for interpreting the applicability of the results and reducing potential biases.
- The HRI community is developing a supportive culture for reproducible research. This includes increased adoption of pre-registration, efforts to release code and bench-marking datasets, as well as providing revised guidelines to the reviewers and editors to encourage acceptance of replication studies and reports on negative results. As an example, the HRI 2025 conference full paper submission guidelines [Johal et al. (2024)] emphasised reproducibility in the technical track, user studies track, and artifact submissions, with reproducibility studies welcomed in the user studies track.
- Due to the complexity in HRI research and development, an iterative process consisting of multiple cycles of design-implement-evaluation-redesign enables researchers to incrementally increase the ecological validity of their findings and refine their research.

Part II

Advanced Topics in HRI Methodology

8

A Framework for Ethical HRI Research

As was discussed in Part I, ethical and responsible conduct should be at the core of HRI research throughout the HRI study life cycle, as with all areas of research that aim to benefit society through technological advancements. In most research settings, the basic infrastructure, such as ethics committees that oversee and approve study designs, as well as the training provided to new researchers on ethical research conduct, is already in place. However, this infrastructure is often adapted from a medical research context not tailored for HRI and may be inadequate to ensure that the values and requirements held by different stakeholder groups are understood and incorporated. This chapter discusses ethical considerations of designing HRI systems and conducting HRI studies as informed by the IEEE Standard for ethical system design [IEEE (2021)]. We propose a framework to support ethical HRI research and experimental practice, which motivates the expanded set of methodologies and practical guidelines for HRI studies covered in the chapters of Part II.

8.1 A Framework for Ethical HRI Research

There have been a number of human research ethics guidelines developed at the international or national levels [OASH (2021), DFAT (2021), NHMRC (2023)]. These guidelines are often developed with medical or social science research as the starting point, while the basic principles can also apply to HRI research with human participants or human data. Recent advancements in AI and automation have exposed an urgent need for frameworks that can safeguard ethical research and application of these emerging technologies. For example, Salmon et al. (2024) conducted a series of workshops to identify ergonomics methods that can assess the adherence of AI technologies used in defence applications with ethical principles. The recent IEEE standard for ethical system design, or [IEEE 7000-2021 IEEE (2021)]

DOI: 10.1201/9781032693682-10

was developed in consultation with experts in system design, ethics, and legislation to identify how intelligent systems can address the ethical values of stakeholders and deliver on both ethical and functional requirements. It outlines considerations during the concept exploration and system development stages to ensure ethical practices. This standard highlights the transition from "useful systems" to systems "designed for good [with] human values embedded".

In this chapter, we summarise five key aspects of IEEE 7000-2021 and discuss them in the context of HRI, highlighting examples of how ethical challenges in HRI research can be considered using the framework. Note that this chapter focuses on designing and conducting ethical HRI research studies, rather than on the ethical application or commercialisation of HRI systems. The IEEE standard's five key steps adapted to the HRI context are as follows:

1. Conceptualise HRI research questions and hypotheses by identifying the stakeholders and use context.

2. Identify ethical values of stakeholders to generate requirements for HRI systems and studies.

3. Validate ethical and functional requirements of HRI systems in cooperation with stakeholders.

4. Beyond measuring benefits, plan for and mitigate risks that may violate the ethical values and requirements.

5. Communicate HRI study design, protocols, analysis, and findings in a transparent manner.

These ethical considerations together with the HRI challenges of interdisciplinarity, context-dependency, and experimental rigour motivate our discussion of adopting an expanded set of human-centred approaches for designing and conducting HRI studies in Part II of this book. Figure 8.1 summarises how these key ethical considerations correspond to the different phases of the HRI study life cycle that were presented in Chapter 1. Note that the proposed framework of ethical HRI studies with human-centred methods is intended as a starting point for HRI researchers to explore these study designs. As shown in Figure 8.1, to achieve ethical HRI studies, there are three main steps, namely:

1. Design an ethical HRI system by addressing stakeholders and context.

2. Evaluate aspects of this HRI system in an ethically designed and conducted HRI experiment study.

3. Analyse and report findings of this HRI study in a transparent manner.

The key ethical considerations in the first step are identifying contexts, stakeholders, and their ethical values, including autonomy, equality, privacy, respect, fairness, and beneficence. The key ethical considerations in the second step are validating ethical and functional requirements and risk identification and reduction. The key ethical considerations in the third step are validating ethical and functional requirements and transparency management.

Step 1 represents activities that occur prior to an HRI study life cycle and focuses on understanding the HRI system in a human-centred and interdisciplinary framework and identifying HRI studies to conduct. Here, the ethical and functional requirements of the HRI system are identified, as we will discuss further in this chapter. Step 2 represents the design and experiment phases of an HRI study life cycle and focuses on conceptualising HRI research questions and hypotheses based on stakeholders and contexts (see Chapter 2), adopting human-centred experiment design and evaluation methods (see Chapter 3), and recruiting participants representative of stakeholders while considering privacy and consent of participants (see Chapter 4). These ethical considerations during the design and experiment phases are further covered in the advanced HRI methodologies that will be presented in Chapters 9 to 12. Step 3 represents the analysis and post-study phases of an HRI study life cycle and focuses on applying data collection and analysis approaches reflective of context and stakeholders' requirements and on considerations in quality, validity, and reliability of data and HRI study findings (see Chapters 5 and 6).

8.2 Identifying Contexts and Stakeholders

IEEE 7000-2021 states that ethical system design "defines how a system is expected to operate from the users' perspective and its context of use, its stakeholders, and its potential for ethical benefit or harm". It is challenging to identify and continuously engage stakeholders in the research process [Blok (2022)]. A responsible governance model requires mutual learning that develops a shared vocabulary between researchers and users, such as "translating" technical terms describing a robot's functions to be more meaningful to people without a technical background [Tian et al. (2021)]. The inclusion of stakeholders relies on transparency and openness during research and reporting, as will be discussed in Section 8.6.

Stakeholders' contexts can be implemented by formulating hypotheses of HRI studies around how people understand and interact with robots and determining what are the benefits and risks of an HRI system in the social, cultural, personal, and other contexts of its application (see Chapter 2).

An example of how the various stakeholders can bring different ethical challenges in HRI can be found in robotics for elder care [Sharkey and Sharkey

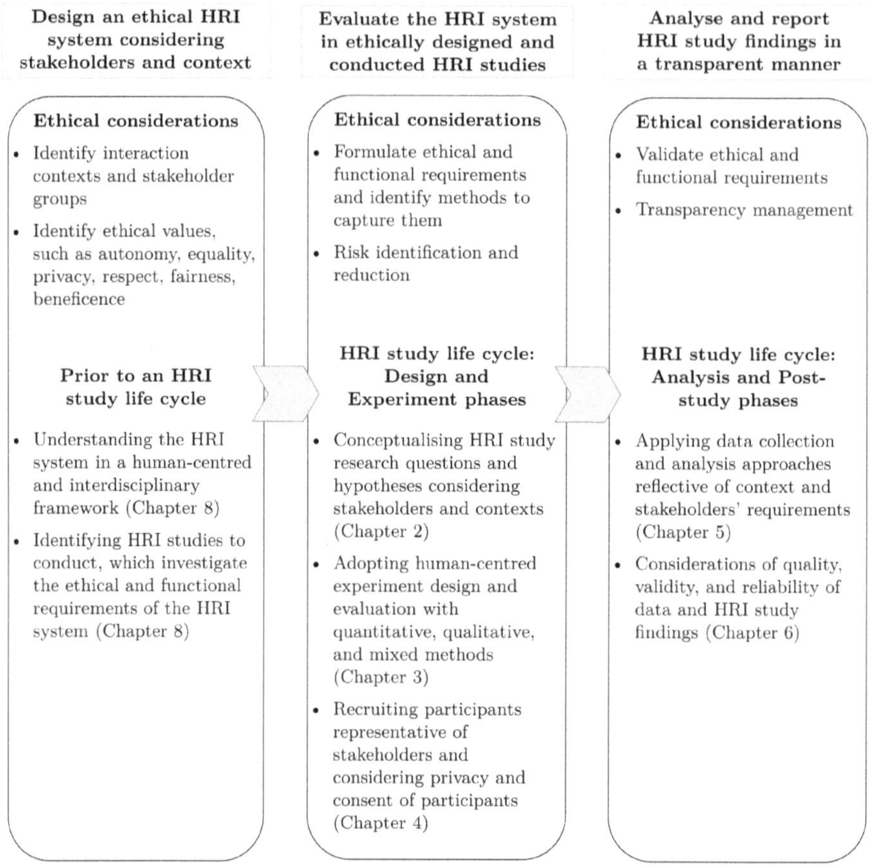

FIGURE 8.1
A framework for ethical HRI studies with human-centred research methodologies.

(2012)]. Only considering the caretakers' requirements may lead to patients feeling objectified, while only considering the patients' requirements may lead to caretakers' distrust towards the robot. The broader interaction and societal contexts, such as the social connection provided by human caretakers beyond the task itself, further influence the potential benefits or risks of HRI systems. Thus, while HRI systems have the potential to address societal needs, they may also cause harm if the values and needs of different stakeholders within the usage context are not fully understood.

In another example, Arevalo Arboleda et al. (2021) developed assistive robots for people with motor disabilities (PWMD) for both the home and work environments. The authors considered PWMD to be the primary users of the robot and caregivers, engineers, social workers, social pedagogists, and workstation supervisors to be the secondary users. Their study demonstrates that involving different stakeholders in the design process supports the identification and prioritisation of different, nuanced, and sometimes conflicting interests of these stakeholders. For example, while some workstation supervisors prioritise production, social workers at the same facility prioritise well-being.

Cespedes et al. (2020) developed assistive robots for rehabilitation. They engaged both clinicians and patients to design a system that incorporated inertial measurement unit (IMU) sensors, a Lokomat (a treadmill that provides body weight support), an NAO robot, and therapists to deliver gait rehabilitation sessions. The IMU sensors allow for a continuous and detailed monitoring of the patient's posture during therapy, while the NAO robot can provide feedback and encouragement. This setup reduced the workload of the human therapists, allowing them to focus on enhancing other aspects of the therapy. In a series of Wizard-of-Oz (WoZ) studies, Leme et al. (2017, 2021) used mobile robots in overground gait training to provide companionship to patients during therapy. These studies show that when the context and user's perspective are properly identified, robots can promote positive social interactions while reducing the burden of the person delivering care. By formulating research questions and study designs driven by stakeholders, HRI research can move towards a more ethical and beneficial outcome.

8.3 Identifying Ethical Values

The development of ethical HRI systems relies on clearly identifying what the ethical values of the stakeholders are, how they prioritise these values, and how these values can be measured. IEEE 7000-2021 recommends an ethical design framework driven by ethical theories for value identification and proposes iterative value prioritisation.

Ethical values for HRI continuously emerge with its advancement and application in society. Values can be identified from three perspectives: *utilitarian*, considering the ethical value of outcomes [Mill and Bentham (1987)]; *virtue ethics*, considering the character and flourishing of agents [MacIntyre (2020)]; and *duty ethics*, considering responsibilities for maintaining shared morals in society [Kant (2013)].

With the stakeholders and context of an HRI system defined, researchers can engage in iterative and participatory design activities, which will be discussed in Chapter 11, to elicit ethical values collaboratively. After a set of values is identified, they may require prioritisation due to feasibility concerns or conflict between individual values. IEEE 7000-2021 recommends ranking ethical values to be more important if they are more persistent, less divisible, independent from other values, less dependent on a specific bearer, and lead to deeper and longer human satisfaction. Continuing with the example from Arevalo Arboleda et al. (2021), the well-being of PWMD is more important than their productivity in the workplace if these two ethical values are in conflict. The well-being leads to better quality of life for PWMD, while higher productivity only contributes to temporary financial gain in the workplace. Here we discuss the five key ethical values relevant for HRI studies, namely, autonomy, equality, privacy, fairness [Boada et al. (2021)], and beneficence.

8.3.1 Autonomy – Impact of HRI on Human-Human and Human-Robot Relationships

In addition to establishing human-robot relationships, HRI introduces new dynamics into human-human relationships. The impacts of these new relationship dynamics need to be evaluated to minimise possible negative effects. For instance, robots with human-like qualities or a high level of autonomy raise concerns regarding their potential to distort the perception of reality (e.g., believing robots are capable of forming friendships and genuine emotions) and their impact on existing human-human relationships (e.g., potentially sacrificing real-world social interactions in favour of interacting with the robot) in the long run [Sparrow and Sparrow (2006)].

The concern that human-human relationships can change when a robot is introduced has also been raised by [Lemaignan et al. (2016)]. The authors argued that social robots can change the dynamics between teachers and students in the education setting. During longitudinal deployment, human-robot interactions can foster human-robot relationships and emotional bonds towards the technology [de Graaf (2016)]. Users may become over-reliant on services provided by a robot, or their behaviours and decisions can be significantly influenced by a robot.

Over-trust and over-reliance can result in automation bias, which reduces the overall performance of the human-robot team [Salem et al. (2015), Parasuraman and Manzey (2010)]. Moreover, this can reduce users' sense of control over the outcomes of robot-assisted tasks and may even impact their own sense of autonomy in general, which can have a severe negative impact on their mental and physical health [Huber et al. (2016), Oviatt (2021)]. An erosion of human autonomy, that is, a weakening of an individual's freedom to exercise agency or control over a task or their behaviours, is a major ethical concern in current adoption of AI and automation [Oviatt (2021)].

A recent work by Rueben et al. (2020b) applied the make-believe theory in psychology to HRI. Under the make-believe framework, robots are "props" that can trigger certain imaginings. Participants' attribution of robots having certain human-like characteristics (e.g., assigning gender or emotional feelings to a robot) is an act of participants imagining a robot's state, which does not necessarily imply they believe a robot to actually have such mental states.

Thus, when conducting HRI studies, the capabilities and limitations of a robot should be communicated to participants and users in a meaningful manner to respect their autonomy and agency.

Acquiring participants' informed consent is important for ensuring their autonomy. However, it is worth noting that this may not always be achievable, especially in HRI research design that involves deception of participants. An example is when robot teleoperation is used to establish a control or experiment condition, while participants are under the impression that the robot is fully autonomous (i.e., WoZ). Another example is when the awareness of the ongoing experiment or observation influences participants' behaviours or perception, that is, the observer's bias [Mahtani et al. (2018)]. Thus covert experimentation has to be conducted with implied or waived participant consent. In these types of research design, it is important to weigh the benefits and risks to the participants, whether or not deception is necessary, and measures that can support participants to maintain or regain autonomy, such as appropriate debriefing after experiments (see Section 4.5.1).

While HRI researchers need to be aware of the aforementioned concerns, previous longitudinal studies suggest that participants' perceptions of robots change throughout the adaptation process, for example, due to habituation or the novelty effect Koay et al. (2007)]. A robot's capabilities also limit the extent of the relationship that a person can form with it. In an example study, Moyle et al. (2017) deployed the PARO zoomorphic robots in long-term care facilities and found that the use of social robots provided only positive benefits to users initially, possibly due to the novelty effect. After benefits decreases, some users stop using the technology or terminated their relationship with the robot. Another longitudinal study tracked the use of the zoomorphic robot Pleo in families with children [Fernaeus et al. (2010)]. They found that participants interacted with the robot out of curiosity and drew comparisons between the robot and pets. However, participants tended to reduce their interaction with Pleo once they realised it had limited capabilities. These studies suggest that the relationship between humans and robots can be managed naturally as users continuously update their expectations. The key is to identify the potential effects and communicate them to users to respect their autonomy and agency.

8.3.2 Equality – Reducing Biases

The ethical and responsible creation, deployment, and use of robots includes avoiding personality traits, tasks, or situations in which robots contribute to the reinforcement of harmful stereotypes. In addition, an imbalanced power relation may be generated by the researcher-participant relationship and differences in technical expertise or perceived efficacy of robot usage, which can be problematic for equality. This may lead to changes in how participants behave or their decision in whether or not to share their honest opinions during an HRI study. The desire to promote equality motivates adopting participatory research methodologies in HRI studies, as will be further discussed in Chapter 11.

Concerns about interactive agents reinforcing biases and stereotypes have been raised in HCI [Benjamin (2021)], AI [Leavy (2018)], and HRI [Sparrow (2020)]. For example, programming feminine voices into robots designed for household domestic tasks and behaviours such as being submissive or flirtatious with its users can continue to reinforce harmful gender stereotypes [Strengers and Kennedy (2020)]. Another example of gender stereotyping is the use of a binary gender framework (male/female) where this information is captured as an independent variable in HRI studies. This does not capture the diversity of gender expression, as discussed in previous research on robot gender [Ghazali et al. (2018)], human gender [Law et al. (2021)], or the interaction of robot and human gender [Jackson et al. (2020)]. Hamidi et al. (2018) conducted a study to examine the effect of gender misrepresentation. In the study, transgender participants were interviewed regarding the use of automatic gender recognition systems, which typically misgender transgender individuals. Participants expressed that it is worse to be misgendered by the automatic system than by humans because the opaque classification algorithm of an automatic system does not allow a user to correct its gender classification errors. Thus, consistent misgendering by an automatic system in long-term interaction reinforces gender dysphoria, which is when transgender people feel discomfort or distressed due to their physical body not matching their gender expression.

Racial discrimination can also exist in the context of robots or HRI. For instance, Bartneck et al. (2018) showed that people can assign race to robots and apply the stereotypical biases accordingly. The authors were able to replicate the shooter bias paradigm (i.e., in first-person shooter task, when participants are shown images of Black or white people holding either a neutral object or a gun, i.e., the targets, Black targets are more readily shot than White targets) using images of racialised NAO robots (i.e., by manipulating the robots' body colour to match the skin tone of a Black or white person). Another concern is using machine learning approaches to automatically customise robot behaviours based on classification of user's expected preferences of appearance [Nomura (2017)], or the use of robots in elder care that adopts recruitment or promotional materials which phrases ageing as a form of disability and paints older adult users as having deficits [Lee et al. (2016)]. These examples again perpetuate biases and undesirable stereotypes.

In a 2021 CHI conference keynote presentation, Professor Ruha Benjamin noted examples of racial biases and stereotypes manifesting in technology design (e.g., using left-justified text alignment as the default design enforces the culture bias of excluding languages with right-to-left scripts) [Benjamin (2021)]. Riek and Howard (2014) also highlighted the lack of diversity in the design of robot appearance, which tends to favour white surfaces or metallic finishes. Recommendations for addressing biases call for a more diversified design [Riek and Howard (2014), Bartneck et al. (2018)], as well as incorporating voices beyond convenience samples with the focus on including participants from diverse cultural, socioeconomical, and racial backgrounds [Benjamin (2021)].

In an example study, Harrington and Dillahunt (2021) conducted a participatory design study that engaged Black youth to envision the future of technology and design. The researchers employed the speculative design methodology to encourage participants' creative thinking (i.e., suspend current beliefs and imagine the future). Another study by Hamidi et al. (2018) explores issues of gender representation in technology. The authors conducted semi-structured interviews with transgender participants to understand their perspectives on automatic gender recognition algorithms. To the participants, the risks of the technology outweighed the benefits, where, similar to the aforementioned studies, the use of technology would reinforce existing oppression. Further, Hamidi et al. (2018), and Scheuerman et al. (2020), Costanza-Chock (2020) discussed design guidelines that encourage gender equity and inclusive practices. For instance, Costanza-Chock (2020) proposed the approach of "design justice" to emphasise the knowledge, contributions, and priorities of marginalised communities in the design process.

8.3.3 Privacy and Respect – Protection against Undesirable Data Collection

Robots often rely on data collection to perform their intended tasks. Understanding and respecting people's expectations and preferences on how their data may be used is important for developing privacy-preserving HRI. For example, Lee et al. (2017) engaged in participatory design with older adults with depression to conceptualise socially assistive robots for supporting them. While the participants mostly requested functionalities that would assist them with physical activities around the house, they did not want any data to be accessible to medical professionals, regardless of the potential usefulness of the data for clinical support purposes. In addition to the data being accessed by clinicians, participants in another study also had concerns about their private activities at home being accessed by unknown parties or the data being wrongfully obtained by insurance companies [Lee et al. (2016)].

To protect privacy of vulnerable populations, a good approach may be to consider the preference of different stakeholders (see Section 8.2). The fact that user data can be collected by a robot does not entitle the designers to have uninhibited access to that data. Support of continuous and transparent human consent and decisions, as suggested in medical research [Kraft et al. (2018)], may be the first step to improve trust between participants and researchers. In HRI studies, these approaches may facilitate more ethical HRI and improve acceptance and trust in robots.

8.3.4 Fairness – Representation and Equitable Access to Technology

HRI studies involving human participants often recruit from "Western, Educated, Industrialised, Rich, and Democratic" (WEIRD) societies, which represent only 12% of the world's population [Henrich et al. (2010)]. Representation of populations is not only an issue in HRI [e.g., Baxter et al. (2016)], but also in HCI [Linxen et al. (2021)] and psychology [Henrich et al. (2010)].

The use of WEIRD participants limits the cultural and socioeconomical diversity of the study sample. A growing number of HRI studies have attempted to encapsulate cultural diversity by recruiting from multiple locations, for example, the United States and China [Wang et al. (2010)] or the United States and Japan [Fraune et al. (2015)]. However, these studies often analyse geographic cultural differences, while generalising that people in the same geographic region hold the same cultural beliefs. Intercultural differences are often overlooked, including issues such as race, gender, ethnicity, and differences in socioeconomical background. Another issue with underrepresented populations in HRI and technology development is limited technology adoption (i.e., technology may be inaccessible or intimidating for a population); the consequences of lack of technology then affects people's proficiency and literacy, forming a negative feedback loop [Harrington and Dillahunt (2021)]. It is also worth noting that the COVID-19 pandemic has exacerbated this negative feedback loop resulting in further detrimental effects: Marginalised populations are excluded from education, healthcare, and employment because they are not able to participate remotely due to a lack of technical proficiency and access to technology. Representation issues in HRI were discussed in Chapter 4 and will be further discussed in Chapter 10.

8.3.5 Beneficence – Maximising Participant Experience and Outcome

Benefits of participating in HRI studies are often small participant honoria, gifts, or early access to a new technology. However, the ability for HRI systems to provide a long-term positive impact to participants is often limited due to the investigative nature of HRI research as well as the prohibitive cost; for example, PARO robots are retrieved post-study despite their known benefits to the participants [Moyle et al. (2019)].

In addition, those partaking in an HRI study may not be the direct beneficiaries of the research outcomes [Kelly (2019)]. The ethical design process elicits further involvement of stakeholders during the design and evaluation of an HRI system. It raises the question of compensating participants for their contributions appropriately, especially when their inputs and intellectual contributions directly lead to design decisions.

Currently, participants often implicitly surrender the ownership to their ideas when they are involved in HRI studies. This may prevent access to full benefits (e.g., financial gains from products developed based on HRI studies), and a better practice for communication and compensation in such studies is needed.

Beyond financial compensation (for which institutional guidelines can be applied), it is recommended that researchers maximise both the experience and outcome of participation [Kelly (2019)]. Identifying outcomes in some HRI studies, especially exploratory studies, may not be feasible until further testing, and such possibility should be communicated to the participants. When tangible outcomes are available, it is important for researchers to clearly demonstrate to the participants their contributions, as opportunities to be altruistic may be beneficial, especially for participants who may not directly benefit from the outcome of the study. As suggested by Kelly (2019), cultivating positive experiences, such as gaining new knowledge and forming new relationships, can also be a benefit of participation. This is similar to the study design philosophy of mutual learning for HRI using participatory design [Ostrowski et al. (2021)]. For further discussion on incentives and compensation in HRI studies, please see Chapter 4.

8.4 Validating Ethical and Functional Requirements

Ethical values of stakeholders lead to ethical and functional requirements that HRI systems need to satisfy. As guided by IEEE 7000-2021, these requirements can be developed through the collaborative effort between HRI researchers and stakeholders; the requirements can then shape the system and experimental design. Involving stakeholders in various research phases can be facilitated by participatory design and other qualitative research methods, which provide feedback that is key to eliciting and measuring these requirements, as will be described in Chapter 11. Validating requirements can also be performed iteratively. HRI experiments often attempt to capture a participant's perception of a robot *during* the experiment, which may not capture what the participant takes away from the interaction. This requires follow-up studies and longitudinal research. While longitudinal participatory design studies may be difficult to implement in HRI, researchers can use *pilot studies*, as was discussed in Section 6.1, to investigate whether or not an HRI study would conform to the original ethical and functional requirements.

Furthermore, measurements of human behaviours and social interactions can contain errors that influence the validation of these requirements [Lazar et al. (2017)]. Errors can arise from the testing conditions, for example, speech recognition errors can negatively impact robot acceptance [Werner (2020)]; the type of measurements and instruments, for example, ambiguous and difficult to answer questions; the participants themselves, for example, participants with higher technical affinity may feel less anxious when interacting with robots [Horstmann and Krämer (2019)]; among other factors [Salkind (2010)], as was discussed in Chapter 3 and will be further discussed in Chapter 9.

8.5 Risk Identification and Reduction

The use of robots and other automation technologies may create unintended risks. For example, as Bainbridge (1983) discussed in the application context, the emphasis on a human performing supervision of a robot's autonomous functions can result in lowered situational awareness when human intervention is required to resolve robot failures. There are increased occurrences of these "ironies in automation" as exemplified in the AI Incident Database [Responsible AI Collaborative (2024)], such as autonomous vehicle accidents. Risk management requires a sociotechnical system with hierarchies at different levels, from the system's in-built mitigation measures to policies and government regulations [Rasmussen (1997)]. In the field of AI, there is increasing research on identifying and mitigating risks [Slattery et al. (2024)]. Identification and mitigation of risks to the ethical values and requirements is also essential in the HRI research context. One approach recommended by IEEE 7000-2021 is to simulate or prototype new functions or features to verify their effectiveness and acceptability, as well as to identify risk reduction approaches.

In HRI studies, the simulation of a new robot function is often achieved through WoZ studies, in which a robot is teleoperated by a human to overcome technical limitations. However, WoZ studies can also create unrealistic expectations, or participants may be left with a false impression of what robots are currently capable of [Riek (2012), Riek and Howard (2014)], leading to the ethical concerns related to human autonomy discussed in Section 8.3. There may be associated risks for the WoZ operators as well. An example is when studies require WoZ operators to perform scenarios such as cheating [Short et al. (2010)], to endure verbal and/or physical abuse, or to simulate other behaviours that would otherwise be socially awkward, such as long periods of silence [Rea et al. (2017)].

Another risk associated with ethical HRI practice is the researchers' ability to influence the results of HRI studies due to limitations in their knowledge and background, being biased towards an experimental condition, or by not properly randomising/stratifying participant groups or fostering relationships with the participants. Researchers may influence the participants, especially in studies where developing relationships with the participants may be a by-product of the study method (e.g., longitudinal studies), or researchers may even be encouraged to foster a relationship with participants as required by the methodology, such as in participatory design. Ostrowski et al. (2021) and Lee et al. (2017) emphasised developing a good rapport with the participants in participatory design studies. While such a relationship can encourage participants to be more engaged, it can also bias them to providing more positive comments towards the robot and the design process.

To reduce such researcher influence, effective approaches include repeating the design step or question over multiple sessions and incorporating opinions of multiple stakeholders [Lee et al. (2017)]. *The Chicago Guide to Collaborative Ethnography* provides further recommendations to researchers related to this issue [Lassiter (2005)], namely:

1. Maintain academic integrity by creating truthful representations of the results.
2. Openly and willingly communicate intentions, plans, goals, and the collaborative processes of the project.
3. Be open to participants' experiences and perspectives, even in the case of conflicts, with all communications using the appropriate language such that the participants can understand.

8.6 Transparency Management

Ethical HRI research requires transparency throughout the conceptualisation, design, experiment, and report stages. During conceptualisation and study design, preregistration, as was discussed in Chapter 6, ensures both positive and null results are published and increases repeatability and reproducibility. It also encourages research results that are less favourable to be published instead of being selectively hidden. Although preregistration does not guarantee true reports, it reduces false positives, hidden confounds, and fraud, by improving transparency [Simmons et al. (2021), Toth et al. (2021)]. In the experimental stage, transparency is maintained by communicating to the stakeholders regarding the study process and their rights (e.g., ability to withdraw). During reporting, transparency means sufficient and appropriate information regarding the ethical and functional values addressed in HRI studies shared both with peer researchers and with stakeholders.

IEEE 7000-2021 recommends researchers publish a clear and transparent account of the data analysis process. This may include pre-study power calculations and effect sizes, data cleaning protocols, results with and without outliers, and negative findings. Publishing other study materials, such as summary data and raw data, analysis scripts, and ethics approval, can benefit the transparency and replication of an HRI study [Szucs and Ioannidis (2017)]. It is also important to report all the steps and decisions made during the data analysis, as well as any researcher-related bias. In qualitative research, researchers are a big part of the research process itself and thus bring their experiences and worldviews in interpreting the data and addressing the research question(s). Thus, reports on the findings need to include direct quotations from the data that are brief and targeted to show the reader the quality of the coding, theme identification and generation, and subsequent conclusions.

Moreover, reducing emphasis on significance testing and encouraging sharing of data, analysis code, and study materials improves transparency, repeatability, and reproducibility. For example, Laban et al. (2021) have made the de-identified dataset, stimuli, and analysis code from their study freely available online, to enable and encourage other researchers to test alternative hypotheses and more exploratory analyses. With online cloud storage platforms and code repositories becoming increasingly available and less costly, sharing of code and datasets has also become more common. However, it is worth noting that participants from whom the data is collected must consent, with full prior disclosure and information, to sharing the data with anyone other than the initial study researchers, as was discussed in the ethical value of privacy in Section 8.3.3.

8.7 Summary

To achieve ethical HRI research practices, beyond basic governance and training in line with common human study ethics codes, researchers conducting HRI studies face the unique challenge of designing and conducting experiments that capture how people interpret and interact with robots. This is different from the traditional engineering-focused approaches that centre around solving technical challenges of autonomous robotics. Engaging direct and indirect users of HRI systems from the beginning of study conceptualisation allows researchers and designers to establish a mutual communication with users on the ethical and functional requirements of an HRI system. Further, involving users throughout the HRI study life cycle instead of only during the evaluation phase as experiment participants can facilitate study designs that are more transparent in measuring the benefits and risks of HRI systems. We recommend that readers review their research questions to identify ethical requirements and suitable validation approaches and review their experimental design to address these ethical considerations.

To gain contextualised insights into users' perception and experience during HRI, which is often subjective and evolving, quantitative methods for empirical HRI studies that analyse factors in isolated, controlled studies may not be sufficient. Thus, Part II will focus on introducing additional qualitative and mixed methods drawn from interdisciplinary research to support designing and conducting ethical HRI research. In Chapters 9 to 12, we will introduce advanced HRI research methodologies for key steps in an HRI study life cycle as expansions to Part I.

Ethical HRI Research Practice

1. Ethical HRI research practice is beyond basic governance and training in line with common human study ethics codes.
2. HRI is human-centred instead of engineering-focused. Users need to be engaged from study conceptualisation and throughout the whole study life cycle.
3. It is important to review research questions and experimental design in order for the research questions and design to effectively validate both functional and ethical requirements.
4. Ensuring ethical outcomes of HRI research requires broader research methodologies with quantitative, qualitative, and mixed methods drawn from interdisciplinary research.

9

Interdisciplinary Approach to HRI Research

HRI requires both advanced technology to generate intelligent and collaborative robots and to understand how people perceive and behave with robots. When formulating research questions, researchers should consider interdisciplinary methods and approaches. This chapter discusses the benefits and challenges of creating and using interdisciplinary approaches to HRI.

9.1 Need for an Interdisciplinary Approach to HRI

Increasingly, HRI research is adopting multidisciplinary theoretical and experimental frameworks, including those from AI, robotics, behavioural and social science, and user experience design. These relevant fields are experiencing a "growing pain", in which some existing practices can reinforce biases or negatively impact individuals, communities, and even the broader society when transitioning from research to applications [Cawthorne (2022), Blok (2022)]. For example, there is a current debate about facial expression recognition technologies trained on biased data and applied in the surveillance context [Tian et al. (2022)]. Similar to other interdisciplinary research, HRI research faces challenges in experimental design and evaluation:

1. *Ethical HRI*: As was discussed in Chapter 8, ethical and responsible conduct is at the core of HRI research, as with all areas of research that aim to benefit society through technological advancements. In most research settings, the basic infrastructure, such as ethics committees that oversee and approve study designs, as well as the training provided to new researchers on ethical research conduct, is already in place. However, this infrastructure is often adapted from a medical research context not tailored for HRI research and may be inadequate to ensure that the values and requirements held by different stakeholder groups are understood and incorporated [Boada et al. (2021)].

DOI: 10.1201/9781032693682-11

2. *Interdisciplinarity*: HRI research relies in parallel on advancing technology to realise intelligent, collaborative robots and understanding how people perceive and behave during interaction with such robots. When formulating research questions, there is a tension between relying on behavioural and social science theories developed for human-human interaction and recognising the differences between human-human interaction and HRI [Leichtmann and Nitsch (2020)]. Furthermore, it is difficult to identify suitable evaluation metrics that capture both task-specific results that address functional requirements and subjective experiences and ethical values of users [Murphy and Schreckenghost (2013)].

3. *Context dependency*: HRI systems are often designed with a particular context in mind, for example, a social robot that helps Dutch preschool children learn English as a foreign language in one-on-one interactions at early learning centres [Vogt et al. (2017)]. The context plays an important role when designing and evaluating HRI systems to ensure that the ethical values and functional requirements of stakeholders are addressed.

4. *Experimental rigour*: High-quality empirical HRI studies that are statistically valid and more broadly generalisable can be challenging to design and implement. The need to increase the transparency and reproducibility of HRI studies is reflected in current reviews of HRI methodologies, for example, measuring perceived robot safety [Lasota et al. (2017)].

In addition, there are open technical and robotics questions about developing effective and safe HRI systems, namely, accurate perception, dexterous manipulation, and safe decision-making [Haddadin (2013), Yanco et al. (2015), Thomaz et al. (2016), Lasota et al. (2017)]. There is a close connection between the technical challenges in robotics and HRI study design, for example, conducting experiments in an empty lab space to avoid confounding factors introduced by a robot's imperfect collision avoidance and navigation function. Addressing such technical challenges is important to advance HRI research. However, we focus on the human experience aspect of HRI.

Taking into account the challenges just mentioned, in Part II of this book, we present an expanded set of research methodologies focusing on a *human-centred* perspective to advance ethical HRI research that leads to beneficial results and positive social impacts, as was summarised in Figure 8.1. Human-centred design

> ...aims to make systems usable and useful by focusing on users, their needs and requirements, and applying knowledge and techniques of human factors/ergonomics and usability. This approach improves effectiveness and efficiency, improves human

> well-being, user satisfaction, accessibility, and sustainability; and counteracts possible adverse effects of use on human health, safety, and performance. [ISO (2019)]

Note that while we introduce several mixed-method and interdisciplinary research methodologies in this book that researchers are encouraged to explore in their HRI studies, our reviews and recommendations are not exhaustive. We encourage HRI researchers to identify collaborators and domain experts that align with their particular study context and with whom they can develop an appropriately tailored interdisciplinary research framework and shared vocabulary.

9.2 Challenges for Interdisciplinary HRI Research

We survey existing reviews and guidelines for HRI research methodology. Our review illustrates the need for an expanded set of study approaches to capture human experiences and behaviours to inform the design of ethical HRI systems and evaluate their results in a contextualised way.

9.2.1 Ethical HRI Research and Applications

An important topic that concerns the HRI community is ensuring ethical HRI research and applications. A recent review analysed 73 studies published in *ACM Transactions on Human-Robot Interaction*, the HRI conference, and the RO-MAN conference [Rosén et al. (2021)]. This review reported on five components of ethical research conduct: Ethics board approval, informed consent, data protection and privacy, deception, and debriefing. Rosén et al. (2021) found that four out of five principles were mentioned in less than one-third of the reviewed publications, with informed consent being the most frequently mentioned principle, reported in 49% of the reviewed papers. The authors advocated that guidelines for ethical research conduct specific to challenges in HRI research are required, especially for deception involving human-like robots and transparent communication of a robot's capabilities and limits to its users. Other ethical issues have also been raised in social assistive robotics, namely, well-being issues (e.g., privacy, autonomy, and loss of human contact), care issues (e.g., the quality of practice, trust, and impact on the concept of care), and justice issues (e.g., legitimacy of the technology, the responsibilities, social equality, and ecological sustainability) [Boada et al. (2021)]. The authors identified that a structured ethical practice framework aimed at benefiting the well-being of the interactors is missing in current social assistive robots research. In the related field of Human-Computer Interaction, Frauenberger et al. (2017) proposed an "In-Action Ethics" framework for exploratory and in-the-wild studies, which offers a situated and responsive ethics approach beyond the commonly adopted anticipatory ethical approval by a review committee. These reviews demonstrate a need for guidelines beyond the current practices, to ensure that HRI systems are designed and evaluated in alignment with the ethical values and requirements of the stakeholders who are directly or indirectly impacted by the HRI systems. Thus, in Section 8.1, we proposed a framework that facilitates ethical HRI research practices through adoption of IEEE 7000-2021 [IEEE (2021)].

9.2.2 Broadening HRI with an Expanded Set of Methodologies

Achieving beneficial outcomes in HRI requires the understanding of how users interact with and perceive a robot in a specific scenario. Thus, HRI researchers often draw on insights from related fields, such as behavioural psychology, social science, or user experience design [Lupetti et al. (2021)]. For example, Mutlu and Forlizzi (2008) demonstrated how robot designs informed by ethnographic analyses led to better integration of the robots in the physical environment, social dynamic, and work flow of the hospital that they were deployed at. Many HRI methodology reviews have focused on improving the rigour and reproducibility of empirical HRI studies, as will be detailed in Section 9.2.4. However, Fischer (2021) suggested that such reviews often overlook the complexity of interaction dynamics. The traditional hypothesis-driven approach may not be suitable for research on emergent and interactive phenomena in HRI. In addition, variables in HRI research are often investigated in isolation, while the combined effect of multiple factors during an interaction is less understood. We are beginning to see more holistic evaluation approaches, such as the USUS framework, which measures the usability, social acceptance, user experience, and societal impact of HRI systems [Weiss et al. (2009)]. Hence, the HRI community can benefit from broadening the range of accepted methodologies that can capture contextual, subjective, and personal aspects of HRI. This may include using qualitative and mixed methods, such as co-design or case studies, which are commonly adopted in artistic, social science, or design fields.

9.2.3 Improving the Contextual Validity of HRI Studies

Many HRI systems are designed with the goal of application to real-life education, health, and elder care scenarios [Dautenhahn (2007)]. To assess the outcomes of such HRI systems, it is critical to evaluate HRI in-situ. However, Zafrani and Nimrod (2019) found that in HRI studies aimed at applications of elder care, two-thirds of the studies had fewer than 30 participants, the majority of which lasted a day, and only 11% lasted 30 days or more, while ideally large-scale longitudinal studies should be implemented to understand the impact of such HRI systems. In addition, hypotheses in HRI studies are often grounded by theories on human communication and interaction under the assumption that robots are treated as social actors in a manner similar to humans. However, the applicability of human-human interaction theories in the HRI context remains unclear. For example, Irfan et al. (2018) presented their failures to replicate the social facilitation theory in HRI and highlighted the challenges of observing and measuring social interaction in HRI. When studying HRI with a target population in real-life settings, it can be impractical to conduct experiments that carry sufficient statistical power due to the limitation in the resources and availability of the participants. However, such HRI studies can lead to novel discoveries and contribute invaluable insight to understanding HRI in-situ. Thus, Seibt et al. (2021) argued that a mixed methodology beyond quantitative analysis can be helpful in capturing the sociocultural context and subjective HRI experience. This understanding is critical for designing socially acceptable and culturally appropriate HRI systems.

9.2.4 Improving Scientific Rigour and Transparency of Empirical HRI Studies

Recently, increased importance has been placed on adopting a rigorous study methodology across multiple scientific fields. For example, behavioural and social science is experiencing a "replication crisis" [Maxwell et al. (2015)], in which efforts to replicate existing studies often do not yield the same observations or conclusions. Replication failure can impact the validity of the proposed findings, leading to incorrect assumptions about the studied phenomena for future work, confusion, and a waste of resources [Ullman et al. (2021)]. To improve the scientific rigour of empirical HRI studies, other reviews focus on identifying reliable metrics for evaluating HRI systems [e.g., Schrum et al. (2020)]. More than a decade ago, Bethel and Murphy (2010) identified study size and lack of multiple assessment methods as the primary issues in existing studies. A recent review on the use of Likert scales, a common approach to measure subjective judgements of participants in HRI studies, found that only 3 out of the 110 papers performed their analyses correctly [Schrum et al. (2020)]. This demonstrates the urgent need for methodological guidelines to ensure the quality and robustness of HRI studies, especially regarding the validity of evaluation methods and appropriate analyses of different metrics. However, experiment rigour that can potentially be achieved through study designs such as randomised controlled trials [examined in Robinson et al. (2019)] may be difficult for HRI studies due to resource accessibility [Baxter et al. (2016)]. In addition to adopting scientifically rigorous methodologies, recent HRI methodology reviews have emphasised transparency in reporting [e.g., Marvel et al. (2020)]. For example, Belhassein et al. (2019) provided recommendations for the experiment and post-study phases of HRI user studies with a focus on recruiting diverse participants and documenting and reporting studies in a transparent manner.

9.3 Summary

In this chapter, we review the existing methodological guidelines for HRI research. This illustrates the need for an advanced experimental methodology that incorporates interdisciplinary approaches to design and evaluate HRI studies.

Recommendations on an Interdisciplinary Approach to HRI

- **Ethical HRI through human-centred approaches**: Collaborating with stakeholders is critical to designing robots that are socially acceptable and beneficial. However, the methodology for considering ethics throughout design and evaluation is not well-established [Rosén et al. (2021), Boada et al. (2021)]. In Section 8.1, we proposed an ethical framework to motivate the human-centred methodologies we introduce in this paper.
- **Incorporating theories and methodologies from multiple disciplines**: Validated theories in psychology and social science can help generate HRI hypotheses. However, there remain significant differences between human-human interaction and HRI that require researchers' attention. Qualitative and mixed methodologies often adopted in these related fields can help HRI researchers gain insights into the interaction dynamics and complexity in addition to more typical quantitative analysis, which will be further discussed in the rest of Part II of this book.
- **Context and ecological validity**: Human perception and behaviours during interaction are context dependent. When possible, identifying the social and interactive context of HRI systems and accounting for them in the study design can produce observations with more ecological validity.
- **Rigour and transparency**: The validity of the evaluation metrics used is directly related to the validity of an HRI study. Using multiple metrics and evaluation approaches to measure a parameter allows HRI researchers to investigate the convergence of these metrics. We recommend that researchers report detailed study design, cover both significant and non-significant results in the discussion, and open-source the code and data used if possible.

10

Advanced Statistical Analysis Methods in Quantitative HRI Research

This chapter presents advanced methods for quantitative HRI research. This includes handling errors and biases in data and statistical analysis methods beyond Null Hypothesis Significance Testing. Due to constraints such as accessibility to robots, experimental sites, or difficulty in recruiting certain participant groups, HRI studies often have a small sample size. This chapter also provides recommendations for analysing quantitative data with small sizes.

10.1 Common Sources of Error and Bias in User Studies

As in the case of HCI studies, HRI studies are challenging because measurements of human behavior and social interactions can potentially suffer from higher fluctuations [Lazar et al. (2017)]. These fluctuations, often referred to as measurement errors, can arise from testing conditions [e.g., in Werner (2020), the robot's speech recognition kept failing during the interaction, which negatively impacted acceptance ratings], the type of measurements and instruments (e.g., questions were ambiguous and difficult to answer), the participants themselves [e.g., in Horstmann and Krämer (2019), participants with a higher technical affinity often had more realistic expectations about robot capabilities and thus felt less anxious when interacting with robots], among other factors [Salkind (2010)].

Errors can be random or systematic. While random errors are due to chance and cause measurement fluctuations in either direction, systematic errors are often due to external, predictable factors and result in measurements being biased. For example, participants might consistently underperform during all trials of a task because of tiredness or nervousness. Possible sources of systematic errors in HRI studies include:

DOI: 10.1201/9781032693682-12

- *Inappropriate, inaccurate, or incorrectly configured measurement instruments.* For example, physiological measurement instruments such as EMG, EGG, or IMU sensors require an accurate placement on the body [Tiberio et al. (2013)].

- *Inappropriate or unclear experimental procedures.* For example, when a within-subjects design is employed, the randomisation of all conditions is critical. Otherwise, conditions tested later may be consistently better or worse than conditions tested earlier due to learning or fatigue effects. In the case of WoZ studies, if the studies are poorly designed, participants' ratings on the perceived intelligence of the robot are likely to be reflections of the intelligence of the human operator rather than of the robot [Weiss and Bartneck (2015)].

- *Characteristics of participants introducing systematic errors into the results of an HRI study.* As previously mentioned in Section 4.2.1, since many HRI studies employ university students as their main pool of participants, often from a computer science or engineering background, as their main pool of participants, it is likely that the measurements will be biased by the high levels of robotics and technological familiarity of the students [Bartneck et al. (2020b)].

- *Non-intended robot errors due to lack of technical robustness and functionality.* Multiple studies have shown that participants' evaluation of their interaction experience with a robot can be strongly influenced by technical issues. For instance, Hancock et al. (2020) observed that robot reliability plays a substantial role in how much trust participants attribute to a robot. Similarly, Werner (2020) report on a study in which a poorly designed robot frustrated participants and, hence, biased their acceptance ratings.

In addition to these sources of systematic errors, there are also interaction errors and failures are often observed when a human and a robot interact with each other. From the human perspective, Reason (1990) divides human errors into slips, lapses, mistakes, and violations. Slips represent the situations when the action is not what was intended (e.g., accidentally pressing the wrong button). Lapses occur as a result of lapses in a user's memory and/or attention (e.g., forgetting to turn the robot off). Mistakes represent the situation when the intention is not appropriate, and consequently, the performed action is wrong. In other words, slips and lapses are execution failures, while mistakes represent planning failures. Violations represent intentional illegitimate actions (e.g., directing the robot to run into a wall). From the robot perspective, errors can be classified into two classes [Honig and Oron-Gilad (2018)]: *Technical failures* and *interaction failures.* Technical failures are caused by problems in either the robot's hardware or software. Interaction failures are caused by the uncertainties in the robot's interaction with the environment, other agents, and humans [Steinbauer (2012)]. These errors, either from humans or robots, can negatively influence the quality of the collected data. Hence, early detection of the possible sources of interaction errors that might happen during an HRI study is of high importance.

Social desirability bias refers to the tendency of a participant to present oneself and one's social context in a way that is perceived to be socially acceptable, but not reflective of one's reality. For detecting this type of bias, researchers mostly rely on their intuition, knowledge, and previous exposure to identify common cues for social desirability tendencies. These cues include denial of (already known) problems, challenges, or shortcomings; providing only partial or vague answers (paltering); nervous facial expression and other body language cues; and using vocabulary that is more aligned with the experimenter's description of the study rather than their own vocabulary [Bergen and Labonté (2020)]. Social desirability bias can be minimised in private one-on-one interviews by consistently ensuring the participants that their data are truly confidential and that the experiment location is private and with proper soundproofing. Furthermore, the wording and delivery of questions by the researcher might steer participants toward biased or unbiased responses (see Section 12.1). In addition, rapport-building techniques such as humour, self-disclosure, and researcher-participant matching by sociodemographic characteristics can contribute to decreasing the social desirability tendency [Bergen and Labonté (2020)].

Regarding researcher bias, Poggenpoel and Myburgh (2003) summarised three potential causes as follows: (1) The researcher's culture and background affect their judgement on the collected data and the information obtained from data analysis, (2) the researcher is not sufficiently prepared to conduct the field research (e.g., not checking the context and culture of the participants beforehand), and (3) the researcher conducts inappropriate interviews (e.g., leading questions and framing bias). A common procedure for detecting potential researcher biases is to conduct a pilot study in which researchers try out their proposed methods to see if the planned procedures perform as envisioned [Chenail (2011)]. To minimise the researcher bias and strengthen the validity of participants' claims, Riddoch and Cross (2021) proposed a combination of methodological approaches (e.g., behavioural and qualitative), using double-blinding procedures and following guidelines regarding researcher behaviour and demeanour [Bergen and Labonté (2020)]. While there is no universally agreed-upon framework for ensuring scientific rigour when using qualitative approaches, several social and health science studies detailed how researchers can improve the rigour of their own qualitative research practice [Braun and Clarke (2006), Nowell et al. (2017)], Sundler et al. (2019)]. Finally, it is recommended that researchers disclose their background when presenting their findings so that the results can be interpreted in that context, as was discussed in Section 8.6.

The impact of random errors can be reduced by increasing the observed sample size or averaging multiple measurements. When dealing with systematic errors, Lazar et al. (2017) recommend that researchers (1) aim to eliminate or control all possible sources of bias during the study and data collection (e.g., fatigue can be improved by controlling the testing time of the day or the duration of the study test sessions) and (2) isolate their impact from the main effect of interest when analysing the data.

When eliminating and/or controlling possible sources of systematic errors, it is important to keep in mind that the correction to be applied depends on the problem encountered. For instance, systematic errors caused by measurement instruments can be controlled by carefully examining and testing all instruments before each experiment session, as well as monitoring all measurements during the experiment. Biases caused by experimental procedures can be reduced through randomised conditions and clear instructions for both participants and experimenters. Finally, running multiple pilot studies before the actual experiment takes place helps to identify and correct biases due to participants and errors in the robot's functionalities.

Recommendation: Although systematic errors cannot be completely eliminated, it is important to properly identify and control them because they represent potential threats to the validity and reliability of the data collected during a user study. Knowledge about potential sources of bias can be very helpful during data cleaning and analysis. Pilot studies can be used to identify potential biases.

10.2 Advanced Statistical Analysis

Null Hypothesis Significance Testing (NHST) remains the statistical method most commonly used in the analysis of quantitative data in confirmatory HRI studies, although widely criticised and questioned in the last decade. It is often expected that HRI researchers employ and report on NHST methods as part of their research activities [Gunes et al. (2022)], which can include the validation of the functional and ethical requirements of the HRI system being evaluated. Within the framework of NHST, the data collected during an HRI study are used to test whether the null hypothesis is tenable or not [Lang et al. (2017)].

Despite its popularity, recently there have been multiple critiques on the methodological limitations of NHST, including proneness to over-estimating effects (even when there are none), the propensity to over-interpretation or misinterpretation by researchers, the tendency to ignore power and side effects, and sensitivity to sample size [Kelter (2020), Szucs and Ioannidis (2017)].

In this section, we review alternatives to NHST including non-parametric methods, robust statistical methods, regression analysis, and Bayesian methods.

10.2.1 Non-Parametric Methods

When deviation from parametric assumptions is a concern, the sample size is small, or there is uncertainty about the type of distribution, non-parametric methods and modern robust statistical tests offer viable alternatives [Little et al. (2013)].

Non-parametric methods require minimal assumptions about the underlying distribution generating the observed data. This is achieved by ranking the original data instead of using the data itself and the assessment of test statistical significance through randomisation tests. Ranks have the advantage of not being affected by the presence of outliers or skewed distributions and allow for test statistics distributions that do not depend on assumptions such as normality. Similarly, randomisation tests, in which all ways to permute the data are considered, allow for assessing a test statistic significance without making explicit assumptions regarding distributions [Freund et al. (2010)]. Some of the most commonly used non-parametric tests are:

- *Wilcoxon Mann-Whitney test*: Also known as the non-parametric analog of the t-test, this method is a rank-based test for comparing two populations on a continuous outcome using independent samples, for example, comparing the weight of males to that of females. Abd et al. (2017) employed the Wilcoxon Mann-Whitney test for comparing the users' trust towards a Baxter robot operated in different modes.

- *Wilcoxon signed-rank test*: The non-parametric equivalent of the paired samples t-test, this method examines whether two samples were drawn from the same population and can be used when comparing two related samples, matched samples, or repeated measurements on a single sample. Mirnig et al. (2017) utilised the Wilcoxon signed-rank test for comparing the frequencies of different social cues during human interaction with a faulty robot. They found that their experimental data deviated from normality using the Kolmogorov–Smirnov test, and therefore, they used a non-parametric test.

- *Kruskal-Wallis test*: The non-parametric analog to a one-way between-subjects ANOVA, this method analyses the population medians of the ranked form data. Harriott et al. (2015) employed the Kruskal-Wallis test for comparing the mental workload in human-human interaction versus in HRI scenarios.

- *Friedman's test*: The non-parametric equivalent of the repeated-measure ANOVA, this test examines data based on its rank properties. Hypothesis testing for the Friedman's test may be expressed by the medians or the average ranks. Walters et al. (2007) used this test to compare the user's comfort rating for different approaching techniques of a robot in a fetch-and-carry task.

In the context of HRI studies, Schrum et al. (2020) advocates for the use of non-parametric tests when analysing categorical data such as Likert scales. Similarly, recent studies seem to indicate an increase in the use of non-parametric tests. For instance, in a user study comparing the effectiveness of different learning aids (e.g., robots, tablets, and traditional teachers) in the teaching of the Latin script [Sandygulova et al. (2020)], the authors opted to use non-parametric tests after confirming that their collected data violated the normality assumption required for an ANOVA analysis. In Gaudiello et al. (2016) and Wang et al. (2020), where all collected data consisted of Likert-type self-assessment data, the authors decided to use non-parametric tests such as the Kruskal-Wallis and Wilcoxon Mann-Whitney tests.

10.2.2 Robust Statistical Methods

Robust methods, also known as modern methods, are defined as statistical methods particularly designed to provide adequate control of type I error rates, increase the likelihood of discovering relevant differences between groups [Erceg-Hurn and Mirosevich (2008)], and deal with issues such as non-symmetric distributions, outliers, and differences in group variances, among others [Wilcox and Rousselet (2018)].

Robust methods replace traditional regression methods (i.e, ordinary least squares), measures of location (i.e., the mean) and measures of association (i.e., Pearson correlation coefficients) with robust alternatives, such as the sample median, trimmed means, or *Kendall* and *Spearman* coefficients combined with bootstrap methods. These alternative measures can be later used to perform hypothesis testing.

Bootstrapping methods provide precise estimates of population distributions by iteratively resampling cases from a set of observed data. They can be used to obtain accurate confidence interval estimates in the presence of outliers or strongly skewed data [Russell and Dean (2000)]. Rank-based methods, such as the rank transform ANOVA-type statistic and Wilcoxon analysis, offer extensions to classic non-parametric methods and are known to produce valid results when analysing data that are non-normally distributed and/or have different variances between groups [Erceg-Hurn and Mirosevich (2008)].

There are still some limitations associated with the use of these robust alternatives. For instance, there is a reduction in the number of degrees of freedom available for statistical tests, and the estimation of sample size required for sufficient statistical power is more complex [Kitchenham et al. (2017)]. We refer the reader to Erceg-Hurn and Mirosevich (2008), Wilcox and Rousselet (2018), and Kitchenham et al. (2017) for a more detailed introduction to these robust alternative methods.

As an example of the application of robust statistical methods, in their study of how attributions and perceptions of trust are affected by the presence of intelligent artificial mediation during human-computer communication, Hohenstein and Jung (2020) observed that most of their participants' ratings followed a skewed distribution. Instead of a standard mean as a measure of central tendency, the authors computed and reported trimmed mean estimates. Bootstrap methods were also used to compute confidence intervals.

In this section, we introduced alternative statistical methods to NHST. It is important to note, however, that NHST remains an adequate analysis framework in certain contexts, such as guiding researchers during the data-modelling process. Furthermore, many of the issues that arise from NHST can be mitigated with careful design of study hypothesis, data, biases, and expectations, as well as the underlying data-generation process being investigated [Lang et al. (2017)].

10.2.3 Statistical Models and Bayesian Inference Methods

Statistical models, such as regression analysis and Bayesian inference methods, are alternative approaches to classic NHST methods. Rather than simply focusing on the rejection of a hypothesis of no effects, statistical models allow researchers to gather evidence on how substantive, competing hypotheses relate to the collected data and the generation process from which the data was obtained [Kruschke and Liddell (2018)]. They also quantify and characterise the interplay between the underlying process of interest (e.g., the performance of a human-robot team) and any potential experimental or contextual factor of interest [Lazar et al. (2017)]. For instance, Bayesian inference methods account for these factors through the choice of informative priors.

Although often underutilised in the HRI literature compared to other behavioural research communities, such as HCI and psychology, we have started seeing the application of statistical models such as linear regression and Bayesian regression analysis in HRI studies. Since statistical models aim at representing the data-generation process as accurately and completely as possible, they require researchers to not only fully articulate their beliefs and assumptions about the data and the data's generation process, but also to clearly communicate their understanding through the structure of the chosen statistical model.

In recent years, cautionary recommendations against the use of classical statistical approaches based on null-hypothesis testing have been frequently issued [Vandekerckhove et al. (2018), Kaptein and Robertson (2012),

Cumming (2014)]. In particular, there has been a wide concern regarding the fact that p-values alone do not convey information about the size of the effect being analysed or the precision with which this effect is estimated. Alongside these warnings, there have been recurring calls to shift from classic frequentist methods, such as null hypothesis testing, to Bayesian inference methods. While hypothesis testing provides a p-value that indicates the probability of a given observation (i.e., an estimate or the difference between groups) is due to chance, Bayesian approaches provide a relative comparison of how well a null hypothesis and an alternative hypothesis account for the actual data [Kruschke and Liddell (2018)]. That is, Bayesian methods allow researchers to make strong observations about the probability of a given phenomenon based on the collected evidence.

Bayesian methods offer multiple advantages, such as robustness in low-power situations (e.g., small sample size) and quantification of uncertainty. Moreover, they allow for the inclusion of relevant context or domain information through the choice of informative priors, and they offer a powerful framework to build and test complex models [Kruschke and Liddell (2018)]. It is important to notice, however, that Bayesian methods require expertise to be used correctly since they are highly dependent on a good model specification [Schad et al. (2019)]. A general recommendation is to consult with an expert, especially when working with continuous data [Dix (2020)]. Furthermore, researchers should be aware that, compared to classical frequentist approaches, Bayesian methods also require larger sample sizes [Brysbaert (2019)].

Although still new, we have started seeing the application of Bayesian methods in the analysis of HRI studies. For instance, while analysing the relation between participants' extroversion and their tendency to view a robot as anthropomorphic, Kaplan et al. (2019) performed a Bayesian regression analysis to determine the model that best predicted the participants' ratings of the robots, based on their personality variables. In Bodala et al. (2020), one-way Bayesian ANOVA was used to confirm whether the workload manipulation proposed by the authors had an effect on the participants' perceived workload as measured by the NASA-TLX questionnaire. Hamilton et al. (2020) employed a Bayesian repeated measures analysis of variance to determine whether different categories of mixed-reality deictic gestures used to augment the gestural capabilities of non-humanoid robots affected task-related metrics such as accuracy and reaction time.

We refer the reader to Kruschke and Liddell (2018), Etz et al. (2018), Ly et al. (2018), and Etz and Vandekerckhove (2018) for a detailed introduction to Bayesian statistics. A concrete and detailed example of how to use such methods can be found in Schad et al. (2019).

10.2.4 Relationship and Regression Analysis

Relationship and regression analysis assumes a linear relationship between the dependent variable and the independent variables (i.e., the predictors). The analysis aims to quantify the linear relationship, that is, how much of the variance observed for the dependent variable is explained by the predictors, and make predictions of the values of the dependent variable based on the observed values of the predictors [Lazar et al. (2017)]. Linear regression models can also be used to measure the strength of the relationship between the dependent variable and the predictors, make inferences about the significance and predictive power of the predictors, and compute prediction and confidence intervals for a given set of values of the independent predictors [Chatterjee and Simonoff (2012)].

Although often underutilised in the HRI literature compared to other behavioural research communities, such as HCI and affective computing, we have started seeing the application of linear regression models and analysis in HRI studies. Traeger et al. (2020) used a multilevel regression analysis to determine how a robot's social behaviours influence the conversation dynamics between the human members of a human-robot group. Ceha et al. (2019) used regression analysis to understand how robot curiosity is perceived by children, whether a question-asking behaviour can lead to curiosity contagion between a robot and a group of children, and how these two factors impact learning. We refer the reader to Best and Wolf (2014), Chatterjee and Simonoff (2012), and Ramachandran and Tsokos (2020) for a detailed introduction to regression analysis.

Sometimes the aim of a user study is not to determine the validity of a hypothesis, but rather to analyse the relationship between a dependent variable of interest (e.g., a task performance metric or the final score of a subjective questionnaire) and a number of independent variables or *predictors* (e.g., the factors being manipulated during a user study). Regression analysis is a statistical method that, under the assumption of the existence of a linear relationship between the dependent variable and the predictors, allows the researcher to quantify this relationship (i.e., determine how much of the variance observed for the dependent variable is explained by the predictors) and make predictions of the values of the dependent variable based on the observed values of the predictors [Lazar et al. (2017)].

Linear regression models are constructed based on the purpose of the statistical analysis [Lazar et al. (2017)]. If the purpose of the linear regression analysis is to find the relationship between the dependent variable and all predictors, the linear model is built by simultaneously considering all predictors. However, if the purpose of the analysis is to build a model that explains the relationship between the dependent variable and each individual predictor, a hierarchical procedure in which predictors are added to the model one at a time is more appropriate.

Linear regression models can also be used to measure the strength of the relationship between the dependent variable and the predictors, make inferences about the significance and predictive power of the predictors, and compute prediction and confidence intervals for a given set of values of the independent predictors [Chatterjee and Simonoff (2012)]. As with other statistical methods, for a linear regression model to be appropriate and valid, in addition to the linearity assumption previously mentioned, the following assumptions about the data being modelled are to be respected [Chatterjee and Simonoff (2012), Ramachandran and Tsokos (2020)]:

- The expected difference between the observed and the fitted values of the dependent variable must be equal to 0. This difference is also known as error.
- The variance of the errors remains constant across all values (homoscedasticity).
- None of the independent variables must be a constant or a linear combination of other predictors (co-linearity).
- The errors associated with one observation are not to be correlated with errors of any other observations (auto-correlation).
- The errors must follow a normal distribution (normality). This assumption is critical when the sample size is small.

Violations of the first assumption can serve as indications that the current linear model has been incorrectly specified and relevant predictors are missing or irrelevant predictors are to be removed. Deviations from the other assumptions (i.e., homoscedasticity, co-linearity, auto-correlation, and normality) can potentially result in biased errors and regression coefficients and thus lead to incorrect inferences and misleading significance tests or confidence intervals [Best and Wolf (2014)].

10.3 Small Sample Size Studies in Quantitative HRI Research

10.3.1 Small Sample Size Studies

A study's statistical power is defined as the probability of detecting a significant effect of the factor being manipulated on the dependent variable being measured if it exists. Statistical power is directly tied to the number of samples (or measurements) being analysed. Thus, the likelihood of detecting small or even moderate effects vanishes as the sample size decreases [Morgan (2017)]. Similarly, the likelihood of detecting an effect when there is none (type I error) or failing to detect an effect (type II error) can be reduced with a greater sample size. As a result, statistical analyses are more reliable when the sample size is large and the subsequent statistical power is greater [Maxwell et al. (2008)]. The general recommendation is to perform a pre-study power analysis to estimate the appropriate number of participants and measurements required to achieve adequate statistical power or accurate parameter estimates [Patten and Newhart (2017), Bethel et al. (2020b)]. Tools like G*Power 3 [Faul et al. (2007)] can be used for this purpose. New methods for determining sample size are continuously being developed in related fields. An example is Bayesian sequential design [Bendtsen (2022)], which dynamically determines the sample size based on updated observations instead of recruiting a fixed number of participants to avoid under- and over-recruiting in health intervention trials.

Experimental precision and statistical power can also be achieved through noise reduction and increases in effect size [Dix (2020)]. Techniques for reducing noise include selecting a within-subjects design, choosing a narrow participant group, and/or using scripted tasks. However, these techniques might sacrifice the generalisability of the result due to the strictly controlled experiment design. On the other hand, adjusting the sensitivity of the study can increase the effect size. This can be done by targeting a user group for whom the researchers believe a maximum effect or difference will be observed or by including distractor or target tasks that deliberately expose the effects of the independent variables of interest in the study. It is important to report the effect size fairly and accurately, especially when the sensitivity to effect sizes is adjusted, to avoid misleading or inaccurate conclusions [Maxwell et al. (2008)].

Lakens (2013) and Funder and Ozer (2019) provided further guidelines on performing power analysis and reporting effect size.

10.3.2 Statistical Analysis for Small Sample Sizes

Morgan (2017) provide the following recommendations for when data from small sample size studies still need to be analysed. First, for categorical data, alternative exact tests, such as the Fisher's test, [McDonald (2009)], are preferred since more common tests (e.g., the chi-squared test) are known to lack accuracy with a small sample size. Second, for continuous, interval data, it is often advised to verify key assumptions such as normality and sphericity before employing classic parametric methods such as t-tests or ANOVA. However, in small sample size data, the tests employed to verify these assumptions are likely to be underpowered and may lead to incorrect conclusions about the validity of these assumptions. In these cases, Welch's extensions to the t-test and ANOVA or non-parametric options, such as the Mann-Whitney or Kruskal-Wallis tests, are recommended. It is important to notice, however, that non-parametric tests are known to have less statistical power than their parametric counterparts. Third and last, when visualising small sample size data, the use of common visualisation methods, such as histograms and box plots, can be misleading and hard to interpret. Scatterplots are suggested as the best choice for showing the distribution and trends of the data in this case [Weissgerber et al. (2015)].

10.4 Examples

Kaplan et al. (2019) employed Bayesian regression analysis to study the relation between participants' extroversion and their tendency to view a robot as anthropomorphic. The authors also compared multiple candidate models to determine which model best predicted the participant's ratings of the robots, based on their personality variables.

On the other hand, Correia et al. (2018) used non-parametric methods for analysing their data after they found that both the normality and the homogeneity assumptions were violated. They used the Shapiro-Wilk test for checking the normality assumption and the Levene's test for checking the homogeneity assumption. They compare between two conditions using the Mann-Whitney-U test. This test is similar to the t-test with which a comparison between two groups can be conducted. For pairwise comparison for three conditions or more, the Kruskal-Wallis-H test can be used, which is analogous to the ANOVA test.

10.5 Summary

To date, NHST remains the dominant statistical analysis method in HRI research. However, in the last few years, we have seen multiple critiques on the methodological limitations of this approach, including proneness to over-estimating effects (even when there are none), the propensity to over-interpretation or misinterpretation by researchers, the tendency to ignore power and size effects, and sensitivity to sample size [Kelter (2020), Szucs and Ioannidis (2017)]. It is important to note, however, that NHST remains a perfectly adequate analysis framework and that many of the issues that arise from NHST can be mitigated if researchers take the time to carefully consider their hypothesis, their data, their biases, and expectations as well as the underlying data-generation process being investigated [Lang et al. (2017)].

As an alternative to NHST, the research community is currently promoting a switch toward alternative approaches such as statistical models. Since statistical models aim at representing the data-generation process as accurately and completely as possible, they require researchers to not only fully articulate their beliefs and assumptions about the data and the data's generation process, but also to communicate their understanding through the structure of the chosen statistical model.

Regarding the choice of method and ethical considerations, quantitative data analysis is at the core of confirmatory HRI studies and the validation of the functional requirements of HRI systems. Thus, it is critical to conduct it in an ethical manner. Specifically, any statistical conclusion drawn from the collected data is dependent on the data being analysed, the method used to analyse the data, and the researcher's decisions during the analysis (e.g., data coding, exclusion, and cleaning or choice of prior distribution) process. Therefore, it is important to clearly report any known or suspected limitation, defect, or bias in the data that can affect the integrity, validity, and reliability of the conclusion drawn from it. For instance, omitting the gender, race, and other demographics of participants from which data was obtained can result in an over-generalisation of the results and effects found during data analysis [Henrich et al. (2010)]. Similarly, it is important to understand the assumptions, implications, and limitations of the statistical procedures being used so as to reduce incorrect or misleading results. Researchers must recognise that methodological shortcomings and errors during data analysis have the potential to translate into issues of social and ethical justice, for example, when policymaking regarding robotic systems is informed by the outcomes of HRI research studies. Finally, an open data and open method approach in which researchers provide a replicable "paper trail" of their analysis process (from raw data to published conclusions) is fundamental for transparency [Gelman (2018)], in addition to providing the research community with the information necessary to make an objective judgement and interpretation of the conclusions and observations being reported.

Recommendations for Advanced Statistical Analysis of HRI Studies

- Although all sources of systematic errors cannot be eliminated, it is important to properly identify and control them because they represent potential threats to the validity and reliability of the data collected during a user study. Knowledge about potential sources of bias can also be very helpful during data cleaning and analysis. Pilot studies are a good manner in which potential biases can be identified.
- Look at the data before jumping to the statistical testing stage. It is recommended that the researcher initially do scatter plots and get an initial idea about the spread of the data, anomalies or extreme values, and the match between the data and the model (i.e., is the data linear?) [Dix (2020)].
- Whether you decided to use classic parametric methods, confidence intervals, non-parametric methods, or Bayesian statistics, make sure you understand the statistics properly [Dix (2020)].
- Despite the potential misuse, there is a better general understanding of classic parametric methods. Researchers are more likely to use them correctly, and readers are more likely to understand the results better. However, it is important to not only report p-values, but also, where possible, to report confidence intervals and effect sizes [Dix (2020)]. Include all the data necessary to re-create your results (means, variances, number of participants, demographics, number of conditions, excluded data points, etc.) [Bartneck et al. (2020b)].

11

Qualitative and Mixed-Methods Evaluation for HRI

Qualitative methods can address some of the critiques of quantitative methods to better understand people's opinions on the most salient features and functionalities that robots should have for more suitable and inclusive HRI. Qualitative methods further allow new hypotheses, design concepts, and insights to be generated directly from people and allow researchers to capture free-flowing thoughts or concerns about the proposed robot use case. Mixed methods combine both qualitative and quantitative methodologies. By using mixed methods when collecting and analysing data, researchers may better uncover and understand the workings of complex systems, drawing on the results of both quantitative and qualitative approaches. This chapter offers an overview of emerging qualitative experiment methods useful for the HRI context and presents how mixed-methods HRI studies can be designed.

11.1 Qualitative Methods

Ethical experiment design and evaluation requires stakeholders to be identified, included, and addressed in the design and implementation of an HRI study. This includes creating study protocols and evaluation methods suitable for addressing the robot end-users and communities' needs, values, and priorities. On the scale of a single HRI study, quantitative and/or qualitative methods can be used to collect data representing the use and uptake of robotic systems, for instance, adopting value-sensitive design methods [Friedman et al. (2017)] to investigate how healthcare robots can respect a patient's dignity [Van Wynsberghe (2020)]. Qualitative methods, such as focus group interviews or participatory design workshops, allow researchers to obtain subjective and detailed stakeholder feedback to ensure that their viewpoints, needs, and values are met, and if not, how these can be addressed in future iterations.

DOI: 10.1201/9781032693682-13

Qualitative methods can address some of the critiques of quantitative methods to better understand people's opinions on the most salient features and functionalities that robots should have for more suitable and inclusive HRI. Qualitative methods further allow new hypotheses, design concepts, and insights to be generated directly from people and allow researchers to capture free-flowing thoughts or concerns about the proposed robot use case [Veling and McGinn (2021)]. This section offers an overview of qualitative methods useful for the HRI context. Common methods used to collect such information include qualitative observation, (semi-structured) interviews, focus groups, generative activities, and reflective and narrative accounts [Veling and McGinn (2021)]. Expanding on Veling and McGinn (2021), we recommend best practices for each of these methods and explore example HRI studies that incorporated them.

11.1.1 Observations

Observations are often made without any intervention from the researcher team, which facilitates the ability to capture a more natural response to the robot, without requiring the participant interacting with the robot to directly quantify their reaction or perception of the robot [Song and Chung (2010)]. For instance, Šabanović et al. (2006) deployed a robot receptionist at a conference and at an institute. They recorded people's interaction with the robot using cameras, creating a less intrusive way to monitor the experiment. In the retail context, De Gauquier et al. (2021) recorded video observations to investigate how the placement of a humanoid robot influenced its engagement with customers at a Belgian chocolate store via entertainment applications. In the elder care context, Chang et al. (2014) deployed the PARO robot in the public area of a nursing home for 3 months to observe naturally occurring interactions of residents, staff, and visitors with the robot. They collected behavioural coding and counts for quantitative analysis, as well as qualitative observation field notes.

There may be difficulties with data collection using traditional methods, such as surveys and interviews, in observational studies, but such an observational design can still be adopted to generate informative data. In controlled laboratory environments, structured observations can produce rigorous and systematic qualitative data, in which researchers have narrowed down a set of topics that they wish to explore, all while allowing the participant the opportunity for discourse. Structured observations can also be used to generate quantitative data [Price et al. (2015)]. For instance, if researchers identified human-robot direct eye contact as a metric for engagement, they can track its occurrences instead of noting down the action. The STROBE guideline proposed by Von Elm et al. (2007) can be used to formulate a rigorous plan for conducting a structured observation.

Previously, researchers have made structured observations of social factors along dimensions such as appearance, demeanour, behaviour, knowledge, social confidence, and other non-specific categories to compare the difference between a robot interviewer compared to a human interviewer [Henkel et al. (2019)]. In the context of evaluating ethical requirements, observation data can be used to assess ethical values, such as fairness, by identifying groups that are more disadvantaged (e.g., the robot would break down more often when interacting with participants from one population). Researchers may first collect the data (e.g., through audiovisual recording) and then determine the categories or themes of observation, similar to the procedure in Henkel et al. (2019) and Čaić et al. (2020). Observers must be trained prior to the experiment. We further discussed observer training in Section 4.5.

In addition to recording behavioural patterns, observations of users' utterances can allow users' mental models of the interaction to be inferred. For example, people change their linguistic choices depending on the other party (i.e., the robot) participating in the communication. By observing aspects of the communication, such as types of words used, interpersonal relationship developed (e.g., choice of pronoun), and complexity of words [Fischer et al. (2013)], researchers can gain a systematic understanding of the participants' affordances of the human-robot communication.

One concern of observational studies is that the observed interactions might be less natural or lack real-world relevance. This is known as the Hawthorne effect, where participants might behave differently than they normally would because they know they are being watched [Bethel et al. (2020b)]. One potential solution is to use hidden observation. However, this may not completely mitigate the bias, as the knowledge of participating in a research study itself can affect the result [Mahtani et al. (2018)]. In addition, hidden observation can conflict with the need for transparency and the obtaining of informed consent for ethical HRI research. Thus, additional mitigation, such as a post-study debriefing, may be required.

In practice, observations are often combined with other design elements; for instance, Henkel et al. (2019) combined observation, interview, and quantitative analysis, while Forlizzi et al. (2004) combined observation, ethnography, and interview. Purely observational studies are often part of an in-the-wild research where it may be difficult to collect other data [e.g., Šabanović et al. (2006) and Bohus et al. (2014)]. For example, Alemi et al. (2017) conducted an observational study in which they installed the NAO robot in a kindergarten class to assist with English lessons for children learning English as a foreign language. The interactions were recorded and evaluated by teachers, psychologists, and a technician regarding anxiety (i.e., showing reluctance to talk to the robot or being nervous around the robot), interaction (i.e., interacting with the robot through talking or touching, and attention-seeking behaviours), motivation (i.e., whether children liked the robot and whether they were responding positively to the robot's actions). Observations in such in-the-wild HRI studies offer evaluative insights on HRI systems in a non-intrusive manner at the cost of higher data post-processing, annotation, and analysis demands for the research team.

11.1.2 Interviews

Interviews collect salient information about a behaviour, robot, or use case. They are not confined to the use of quantitative scales (see Section 5.2.5 on self-assessment surveys), which can sometimes miss details or constrict people to respond to stimuli in a defined way. In hypothesis-driven research, interviews can be used to provide additional information to explain the quantitative data [Veling and McGinn (2021)]. Qualitative interview responses can also be used to find new research questions and correlations between response answers and new theories. Interview styles can take on many other forms depending on the intended outcome of the interview, as well as the format, structure, and flow, such as informal conversational interviews, interview guide approaches, standardised open-ended or semi-structured interviews, and closed quantitative interviews [Johnson et al. (2016), Mills and Birks (2014)]. Semi-structured interviews are the most common in HRI studies, which involve pre-defined structured interview questions to begin, followed by unstructured follow-up questions [Veling and McGinn (2021)]. Informal interviews have also been used to check in with participants in longitudinal HRI studies [e.g., Chang and Šabanović (2015)]. For instance, Robinson and Kavanagh (2021) conducted a qualitative interview after a social robot-powered health intervention to identify important social factors, design choices, and perceptions of robots that can be replicated and what to modify or remove in future trials.

- *Informal conversational interviews* are conducted in a casual setting where questions arise from a natural conversation. This means the discussion topics and wording of the questions are not pre-determined and the questions can be personalised. The informal interview structure can be used to check in with participants in longitudinal HRI studies [e.g., Chang and Šabanović (2015)].
- *Interview guide approaches* involve researchers using a pre-defined topic outline and topic set. Question wording is not determined in advance, allowing for more flexible word choice and sequence of topics to make the interview more personalised and relevant. One major concern is that topic or sequence changes may elicit different responses from individuals, meaning that the provided information may not be systematic [Johnson (2016), Loftus and Palmer (1996)].
- *Standardised open-ended interviews and semi-structured interviews* both involve asking participants a list of fixed questions. Semi-structured interviews include spontaneous follow-up questions after the fixed

questions. The fixed questions provide a way to compare participant responses, whereas the follow-up questions provide further insight into the response. Semi-structured interviews are common in HRI studies [e.g., Harris and Sharlin (2011), Hoffman (2019b), Wang et al. (2010)].

- *Closed quantitative interviews* [e.g., de Graaf et al. (2016)] are interviews in which questions and responses are fixed in advance. Interview data collected using closed quantitative interviews are the easiest to analyse, but provide less flexibility in participants' responses, or limit opportunities for the participants to report what they consider as relevant [Johnson et al. (2016)].

One important aspect to explore during HRI interviews is the positive and negative responses that arise from the interaction. Rueben et al. (2021) and Bethel et al. (2020a) both provide guidelines and concrete prompts that may be used when conducting HRI interviews. For instance, Lee et al. (2012) deployed a snack-delivering robot in the wild over 4 months and examined if personalised interactions would improve rapport between the robot and participants. In post-study interviews, researchers gained insight into perceived benefits of personalisation, such as receiving personal attention from the robot and encouraging participants to be more engaged in the interaction. However, personalisation also unintentionally made the participants uncomfortable (e.g., the robot simply stating the participants' choice of snacks may result in the feeling of being judged). In addition, the general experimental setup, where the robot was used to deliver snacks to a workplace, made people feel like they had to be social and polite with the robot because bystanders could see and hear the interaction.

Interviews provide a wealth of information about the interaction from the participants' perspective. One key advantage of adopting a more flexible, open-ended interview structure is the ability to guide the interview towards a direction that respects the participant's autonomy and privacy. By allowing participants to drive the topics of discussion, researchers "[forfeit] power to" participants, equalising the researcher-participant relationship [Hoffmann (2007)], as was discussed in Section 8.3.2. Open-ended interview structures also empower the participants because they are free to discuss topics or avoid topics they do not wish to address [Hoffman (2007)].

However, one potential shortcoming of interviews is that data quality depends on participants reflecting on their experience and clearly explaining thoughts and feelings to the interviewer, which can be difficult for some [Mills and Birks (2014), Bethel et al. (2020b)]. The quality of the interview data is influenced by factors such as the participant's response style, including response acquiescence (yes-saying), response deviation (nay-saying), and social desirability (providing what the participant perceives as appropriate answers) [Bethel et al. (2020b)]. Volunteer participants may respond differently to non-volunteers, as they might inherently be more interested in robots and technology [Patten and Newhart (2017)]. The interview response data may be biased when the bulk of the results is provided by a small percentage of participants.

The interviewer can also unconsciously bias a participant's response about the robot or its role through how they phrase a question, for example, using the phrasing "please tell me what you *like* about this robot" rather than a more neutral phrasing such as "please tell me what you *think* of this robot". The phrasing in first question phrasing may bias a participant to focus on reporting positive perceptions, whereas the latter allows participants to freely report both positive and negative perceptions. When formulating prompts, one starting point is to summarise the participant's last statements and then encourage the participant to talk more about those statements [Holloway and Galvin (2016)]. Neutral phrases, such as "what was the experience like", "how did you feel about ...", "can you tell me more", can be used to formulate exploratory questions [Holloway and Galvin (2016)].

It is important to consider that reported data can be biased if results highlight a small percentage of participants. Thus, it is recommended that the final data report include all perspectives of the participants, both negative and positive. For example, self-selected volunteers might respond differently to those who did not actively choose to engage with the robot, with the self-report group more interested in robots and new technology in general, leading to more positive responses [Patten and Newhart (2017)]. Lastly, consistency between unstructured interviews is difficult to maintain between experiments, thus each qualitative analysis should be taken within its presented context.

Data collected in informal interviews can be difficult to analyse. Researchers looking to conduct qualitative methodologies are encouraged to review papers on the analysis procedure. Examples include Elo and Kyngäs (2008), Smith and Shinebourne (2012), and Vaismoradi et al. (2013a). Qualitative data analysis can be conducted on programs such as NVivo, ATLAS.ti, and other natural language packages available online [Costa et al. (2017)], which will also be discussed in Section 12.1.

In summary, interviews provide researchers ways to directly obtain information from participants. This is useful when researchers are interested in developing a deeper understanding of participants' responses during the experiment. Data gathered through interviews can be influenced by the questions asked, how the interviewer reacts to their responses, and how participants get along with the interviewer. Best practice is to use a member of the research team who is not invested in the final outcome of the study to conduct the interviews to avoid unconscious bias towards reporting experiences that are more favourable to the intended outcome. This can also assist in getting more realistic responses. If possible, interviews should be used in conjunction with other measurements to better quantify the experience.

11.1.3 Focus Groups

Focus groups are group interviews where information is exchanged between the group members and facilitators. Focus groups are effective for eliciting multiple perspectives, gaining consensus, and understanding social context from end-users and stakeholders. For instance, Krupp et al. (2017) conducted three focus group sessions with non-academic participants to study how people from the general community view the impact of telepresence robots on their privacy. As focus groups often lack pre-defined structures, it is more common to use this methodology for generating new designs or theories instead of hypothesis testing [Veling and McGinn (2021)]. Since a large amount of data can be collected in one session, focus groups are considered cost-efficient [Onwuegbuzie et al. (2010)]. The group setting may increase the number of participants willing to sign up and encourage marginalised populations to speak up [Veling and McGinn (2021)]. Similarly, Onwuegbuzie et al. (2010) suggested that focus groups can create a safer environment than individual interviews.

One concern of using focus groups for HRI research is that it can be less suited for discussing sensitive topics and should not be used for discussing personal topics, such as if people would be willing to use a robot to treat their mental illness or for a personal medical treatment [Veling and McGinn (2021), Fusch et al. (2017)]. When designing an HRI focus group, researchers may consider the following aspects summarised from Onwuegbuzie et al. (2010):

- *Size of the group or subgroup*: Having a group of 6–12 participants is recommended, while "mini focus groups" consisting of 3 or 4 participants can be used for gaining in-depth information.
- *Session duration and number*: It is recommended that a focus group meeting be 1 to 2 hours long. Researchers can either perform a single session or a series of three to six sessions depending on the research questions and design.
- *Roles of researchers*: It is recommended that a focus group session have a facilitator who leads the discussion and an assistant facilitator who handles note-taking and other logistic tasks (e.g., recording and interruption management).

Another important consideration is to monitor group dynamics. For instance, if the majority of members are similar to each other, then they are less likely to have differing opinions and others who are not similar to the majority may not express their views and instead feel pressured to conform to the group consensus. Researchers may have more productive and informative sessions if these group dynamics can be identified early by keeping track of utterances and the response distribution [Onwuegbuzie et al. (2010)]. Onwuegbuzie et al. (2010) proposed that these dynamics be monitored through charts and diagrams to systematically collect information on the evolving group dynamics, which may also be beneficial for HRI studies. Finally, focus group data analysis is similar to interviews, where the sessions can be recorded, transcribed, and coded for thematic or content analysis (see Section 12.1 for further discussion on data analysis).

11.1.4 Generative Activities

Generative activities are designed to identify problems, determine possible solutions, or collect information about future steps. In the context of HRI studies, participants can be involved in generative activities to help conceptualise research directions, collect information to understand desirable robot features, or identify deployment opportunities. Example activities include drawing [Lee et al. (2017)]; arts and crafts [Sung et al. (2009), Ostrowski et al. (2021)]; writing [Lee et al. (2017)]; prototyping, for example, providing hands-on experience with sensors or programming robots [Ostrowski et al. (2021), DiSalvo et al. (2012)]; or working with "open-ended objects, for example, a bag of miniature figures, electronic toys, sensors, robot figurines, a massage tool, a Roomba vacuum, and more [Caleb-Solly et al. (2014)]. Careful selection of the activities or objects may elicit different ideas and feelings, so it is important to clearly identify what type of information the researcher is looking to collect in the context of an HRI outcome. The open-ended objects in Caleb-Solly et al. (2014) were specifically selected to provide various tactile feedback to simulate concepts with different embodiment. Beyond prototyping activities, discussion and concept generation can happen through games. For instance, playing a game that involves a researcher-participant pair to envision a new robot design by alternating placing of a game piece that represents the robot's function on the board. Brandt (2006) described these design games in more detail. Generative activities are typically conducted in conjunction with surveys, interviews, and/or focus groups and can be deployed in ethnographic studies, participatory design and co-design [Veling and McGinn (2021)], and human-centred design.

11.1.5 Reflective Narrative

Reflective narrative, such as through the use of diaries, drawings, photos, and videos, are a form of qualitative analysis. It is seen more often in exploratory and design studies [Veling and McGinn (2021)]. This methodology is used to track feelings and perception changes in the absence of the researchers. For instance, Fernaeus et al. (2010) encouraged families with children to take videos and photos during the family's (typically the children's) interaction with the robot toy, Pleo, to supplement two check-in interviews over 2–10 months. Frennert et al. (2017) asked elderly participants to document their interaction with the home assistant HOBBIT robot using a camera and a diary over 3 weeks. Participants were asked to reflect on a short list of questions about the frequency and functionality of the robot daily. Potential issues with reflective and narrative accounts is the reliance on self-initiative to record the experience. As Fernaeus et al. (2010) noted, the amount of data that participants record can decrease drastically after the start, and therefore, it can be difficult to use as a primary data source. However, reflective narrative accounts can be powerful supplementary materials that allow researchers to spot the discrepancy between what participants said versus what they actually did. For instance, researchers noted a discrepancy between how participants described the robot in the interviews (i.e., participants would discuss the robot in terms of future potentials) while not actually using the robot due to it being unable to complete assigned tasks [Frennert et al. (2017)].

11.1.6 Ethnography and Participatory Ethnography Studies

Ethnography and participatory ethnography studies stem from the field of anthropology, with the aim to understand human behaviours within the appropriate social and cultural context. With respect to HRI, researchers aim to understand the feelings, beliefs, and meanings of the relationships between people, as well as the human-robot relationship, within the culture being studied [Fusch et al. (2017)]. Ethnography utilises techniques such as observation, surveys, and interviews to collect qualitative data, with the key idea that data collection happens in-situ and over a long period of time. Most importantly, unlike observational studies where researchers are passively observing, ethnography studies may involve researchers actively participating in the culture or phenomenon while making observations (i.e., being a "participant observer") [Bernard (2017)]. For instance, Blond (2019) conducted an ethnography study over 9 years to examine the adaptation of an elder care robot, Silbot, through the initial testing to the robot's deployment in Korea, Finland, and Denmark. Silbot worked alongside a human game instructor to host brain fitness games for patients with dementia. During the study, the author repeatedly interviewed many stakeholders, attended the demonstration sessions at the municipality, and participated as a participant observer in game sessions with senior citizens. Through this longitudinal ethnography study, Blond (2019) was able to observe the emergent behaviours from both the senior citizens and the human instructors that help explain and justify the robot's behaviours to the senior citizens and the instructors, increasing the interpretability of the same robot functions, allowing them to explore new use cases of the robot, which resulted in better perception and acceptance from the senior citizens and the instructors. This process helped validate existing theories where the context of the robot usage actively shapes the users' acceptance towards and interaction with the robot. Hasse et al. (2019) provides more examples of ethnography studies in HRI.

Ethnography studies allow researchers to better anchor the cultural phenomenons through first-hand experience. This improves the external and internal validity of the observations by reducing observer bias, the novelty effect, and the Hawthorne effect [Bernard (2017)]. By being part of the culture, researchers may also avoid the "sucker bias", where deceiving outsiders is considered appropriate in some cultures [Bernard (2017)]. A successful ethnography study requires researchers to develop "explicit awareness", the ability to notice seemingly mundane phenomena, and strengthen their memory (e.g., remember things long enough to make notes). Bernard (2017) outlines the procedure and techniques researchers may adopt to conduct a productive ethnographic study. Jacobs et al. (2020) provided guidelines targeting ethnography in HRI. Most importantly, it is necessary for researchers to take a non-judgemental stance when conducting ethnographic studies, respecting the cultural and social context.

11.1.7 Participatory Design (PD) and Co-Design

Participatory design (PD) and co-design are commonly used in exploratory studies. Participants are included in early stages of the research process in PD, contributing their knowledge and expertise that can potentially generate new hypotheses and designs [Lee et al. (2017)]. Both PD and co-design emphasise involving people outside of the research team when designing HRI systems and research studies.

PD focuses on the end-user participation, while co-design includes a wider range of stakeholders [Cristol et al. (2020)]. Another difference is the study structure, where co-design involves all parties in joint decision making across all steps, whereas in PD, participants would be involved in a pre-defined set of activities that were designed and facilitated by the researchers. Thus, co-design can help to reduce researcher biases (as was discussed further in Section 10.1). Majority of current HRI studies adopted PD instead of co-design. However, the terms PD and co-design are often used interchangeably when multiple user groups are involved and they modify the designs of robot or the HRI during the activities. Thus, in the remainder of this section, we do not explicitly differentiate PD and co-design.

TABLE 11.1
Participatory Design Phases

What	Where	Who
Initial Exploration Phase		
Researchers meet with participants or stakeholders to understand the context (such as the environment, existing routines and workflow, and experiences) through interviews, observations, and walk-throughs.	Typically on-site (e.g., the end-users' residences)	Individual participants or stakeholders
Discovery Processes Phase		
Researchers utilise generative activities, reflective accounts, presentations, interviews, focus groups, and hands-on experience interacting with the robot to understand the user's values, perceptions, and priorities and allow users to imagine the future.	A convenient space for the participants (e.g., community center or their home)	Multiple participants or stakeholders in a session or individual sessions
Prototyping Phase		
Researchers and participants iteratively refine design through generative activities, or other hands-on activities combined with focus groups, and interviews, while reflecting on the past stages, future usage, and other concerns.	A convenient space for the participants	Multiple participants or stakeholders

PD design involves three basic stages: The initial exploration, discovery processes, and prototyping [Spinuzzi (2005)]. The goal of each phase is highlighted in Table 11.1.

Recent PD methods emphasise mutual learning, where researchers present users with knowledge of the new technology through presentations, live-in experiences, and hands-on prototyping sessions that allow users to actively participate in the development and evaluation [Lee et al. (2017), Ostrowski et al. (2021), Tian et al. (2021)]. For instance, Lee et al. (2017) hosted hands-on prototyping sessions with sensors that a home robot could use. Ostrowski et al. (2021) incorporated both a live-in experience with a robot for a month and programming sessions to design robot behaviours. The activities in a PD study aim to elicit convergent and divergent thinking, where a divergent process promotes ideation (e.g., understanding the participant's perception of robots through drawing and writing) and a convergent process narrows down the focus (e.g., prototyping a robot's behaviour) [Ostrowski et al. (2021)]. Extending the idea of mutual learning, PD studies "continually bring the analysis back to the domain and share it with the participants", meaning participants have the opportunity to perform qualitative analysis together with the researchers to facilitate further discussion [Hansen et al. (2019)].

In addition to the HRI PD guideline that Ostrowski et al. (2021) developed, it is important to highlight the broader societal considerations discussed in Bødker and Kyng (2018). Namely, addressing the "big issues" that will influence future design and adoption of robots with PD. For one, providing impactful changes could mean focusing on technology integration (i.e., influencing the overall workflow) rather than on specific technological advancements [Spinuzzi (2005)]. Moreover, researchers may consider starting institutional changes by engaging the appropriate government agencies [Bødker et al. (2017)]. More detailed guidelines on conducting an effective PD study can be found in Bødker et al. (2017) and [Bødker and Kyng (2018)]; an HRI-specific framework is also introduced in [Axelsson et al. (2021)]. Rooted in ethnography, successful PD is limited by the researcher's ability to make insightful observations. Thus, HRI researchers are encouraged to seek a multidisciplinary team to ensure experimental rigour [Spinuzzi (2005)]. In addition to the research team's expertise, some practical challenges for PD studies include the availability of time, resources, participants' long-term involvement, and institutional commitment [Spinuzzi (2005)]. Further, it is important to acknowledge participants' intellectual contributions and maintain transparent communication with the participants regarding whether or not they will benefit from future academic or commercial outcomes that result from the PD to avoid potentially exploiting participants [Costanza-Chock (2020)].

11.1.8 Case Studies

Case studies are tailored, comprehensive investigations of a new phenomenon within a single study context. A classification of the types of case studies is summarised in the following, based on Baškarada (2014):

- *Exploratory*: For future hypothesis generation
- *Illustrative*: For building a detailed review of an example
- *Critical instance*: For studying a significant (e.g., best or worst) case, or a sanity check during development
- *Program implementation*: For investigating operations across different sites
- *Program effects*: For studying the causality through various assessments
- *Cumulative*: For conducting an evaluation based on multiple case studies

Designing a case study involves case selection, sampling time, coverage, and methods of data collection, as detailed in [Meyer (2001) and Baškarada (2014)]. Case studies can be used as a qualitative methodology [e.g., Lemaignan et al. (2016) and Robins and Dautenhahn (2006)], as well as applied in a mix of qualitative and quantitative methods [Feagin et al. (1991)], [e.g., Oliff et al. (2020) and Khan and Germak (2018)]. Khan and Germak (2018) is an illustrative example of an HRI case study, where the authors investigated if applying design principles from user experience and Human-Computer Interaction would improve the usability of a robot. Researchers conducting case studies need to understand what evidence to collect, the expected variation in data, and what evidence to be considered to support or contradict the theory. Defining the expected behaviour, potentially through the use of pilot studies, can help researchers quickly adjust the data-collection procedure to ensure the appropriate information is collected [Baškarada (2014)]. Qualitative case study design originates from ethnography (hence, the use of interviews and observations); interested readers can refer to Section 11.1.6, which discusses ethnography, for more information.

We have an overview of qualitative methods in the design and evaluation of HRI studies to support the identification and validation of ethical values, stakeholder impacts, and functional outcomes of an HRI system in this section. Quantitative methods provide an objective assessment of how specific factors in HRI influence a population. Qualitative methods provide insights into the subjective experiences and context of individuals. Combining these methods provides a complementary understanding on the benefits and risks of HRI systems. On the scale of a single HRI study, it may be difficult to adopt multiple evaluation methods all at once, and researchers are encouraged to choose the most suitable methods depending on their research questions, study feasibility, and strengths and weaknesses of each method.

11.2 Mixed Methods

Mixed methods combines both the qualitative and quantitative methodologies that were non-exhaustively covered in the previous sections. By using mixed methods when collecting and analysing data, researchers may better uncover and understand the workings of complex systems, drawing on the results of both quantitative and qualitative approaches [Shorten and Smith (2017)]. Seibt et al. (2021) advocated using mixed methods in HRI studies to interpret the developing mental models that people formulate during the interaction, to better capture the perspectival aspect of the HRI experience (e.g., cultural and social contexts), and finally to better inform how current HRI results may ultimately impact future sociocultural norms. Mixed methods can be classified by three characteristics: Level of mixing (partially or fully applied in a study), timing (concurrent or sequential), and approach prioritisation (equal or dominant) [Leech and Onwuegbuzie (2009)]. Eight types of mixed methods can arise from combinations of the characteristics. Here we highlight the methodologies that have been applied in HRI.

A study is known as an *explanatory* sequential mixed-methods design when quantitative data are first collected and analysed and qualitative data are collected to explain the quantitative data [Shorten and Smith (2017)]. Another type, known as *exploratory* sequential mixed-methods design, involves first collecting and analysing qualitative data to form new hypotheses and then collecting quantitative data for theory testing. An HRI example is Choi et al. (2020), where the authors aimed to discover metrics to evaluate the service quality of hospitality robots. A focus group was first conducted and thematically analysed to generate evaluation criteria. A follow-up comparison study on Amazon's MTurk was designed to evaluate the performance of the hospitality robot versus a human staff member, using the criteria developed in the previous stage. Another set of mixed methods is *parallel* mixed-methods design (quantitative and qualitative data are collected and analysed together) and *nested* mixed-methods design (simultaneous collection and analysis, but one method is more dominant). For instance, Wu et al. (2014) conducted a parallel mixed-methods study where they examined the acceptance of an assistive robot, Kompai, by older adults, both healthy and with mild cognitive impairments. Wu et al. (2014) utilised questionnaires, usability-performance measures (e.g., task completion time and number of errors and aids needed), interviews, and focus groups. The quantitative metrics in the study (i.e., performance measures and questionnaire results) allowed the authors to conclude that there is no statistically significant difference between the groups. However, by conducting a thematic analysis, they were able to use qualitative data to complement the quantitative metrics. Furthermore, the authors were able to identify through interviews that participants with cognitive impairments found the system unusable and the experience with the robot discouraging. As demonstrated in these examples, mixed methods facilitates the evaluation of contextualised outcomes of HRI systems, as well as exploratory and confirmatory HRI studies.

Čaić et al. (2020) adopted a mixed-methods study design to compare the social perception (i.e., warmth and competence) of a human coach and a robot coach among elderly participants in a healthcare setting. While quantitative data showed the human coach had a higher social perception, qualitative analysis showed a positive perception of the robot coach, which the authors captured as "it's a machine, but it's kind and gentle". Mixing quantitative and qualitative data facilitates hypothesis testing and comparing findings through narratives.

11.3 Summary

To measure the ethical and functional outcomes of HRI systems, various evaluation and analysis methods can be employed. In Chapter 5, we focused on quantitative methods. In this chapter, we focused on qualitative evaluation and mixed methods. Qualitative methods and mixed methods can provide in-depth narrations or detailed observations about factors that are most salient for people. It also allows for unstructured responses, such as associating HRI with past experiences, emotions, or stories. However, qualitative methods can be difficult to interpret and require a skilled interviewer or facilitator to ask questions in a neutral and informative way. Our overview offers guidance for HRI researchers to adapt qualitative methods to be able to analyse the results with an in-depth and contextualised approach. In Chapter 12, we will further discuss how qualitative data can be analysed.

Recommendations for Qualitative and Mixed-Methods Evaluation

- **Interviews** are useful when researchers are interested in developing a deeper understanding of participants' responses during the experiment. Best practice is to use a member of the research team who is not invested in the final outcome of the study to conduct the interview to avoid unconscious bias towards reporting of experiences that are more favourable towards the intended outcome. If possible, use interviews alongside other measurements to better quantify the experience.

- **Focus groups** help organise and collect a large sample of verbal responses from people, allowing them to build on each other's answers. Focus groups can be challenging to facilitate if there are strong conflicting opinions or personalities in the group, including if some participants dominate or withdraw from the session. These must also be led by a skilled interviewer to avoid bias.

- **Generative activities, ethnographic studies, and reflective narrative accounts** can collect information about HRI in short- and long-term processes, as well as how people's attitudes change over time. Participatory design involves stakeholders early on, which can lead to greater acceptance of the end-product. However, participatory design is resource-intensive, and it may not be suitable when researchers already have an extensive background working with the target sample.

- **Participatory design (PD) and co-design** include users and stakeholders from early stages of the research process. This empowers users to contribute their knowledge and expertise that can potentially generate new hypotheses and designs.
- **Case studies** are tailored, comprehensive investigations of a new phenomenon within a single study context. Understanding what evidence to collect, the expected variation in data, and what evidence to be considered to support or contradict the theory allows the design of more effective case studies.
- We recommend that researchers select evaluation methods most suited to their research questions and address the shortcomings of the selected methods in the discussion. In addition, we recommend that researchers take steps to minimise the risk of bias or influence in the method, which was further discussed in Section 10.1.

12

Qualitative Data and Mixed-Methods Data Analysis

This chapter presents how data collected via qualitative and mixed methods, as was discussed in Chapter 11, can be analysed to provide detailed insights or generate new research directions.

We have introduced various HRI research methods in Parts I and II. In Part III we will provide case studies that illustrate how some of these methods have been adopted in existing research.

12.1 Qualitative Data

Qualitative data analysis is generally described as a non-linear, iterative process [Lester et al. (2020)]. There are five proposed phases for qualitative data cleaning and preparation:

1. *Preparing and organising the data for analysis* is when researchers gather the audio- or video-recorded interview files, convert observational notes to electronic format, and scan documents retrieved in paper form. This phase sets up the collected data in preparation for analysis with digital qualitative data analysis tools, for example, ATLAS.ti or MAXQDA.

2. *Transcribing the data* applies to several types of transcripts used in qualitative research (e.g., multimodal transcript, gisted transcript, verbatim transcript), which is closely related to the methodology employed and the objectives of the research. In addition to manual transcription using tools such as ELAN [Wittenburg et al. (2006)], there are a growing number of automated services (e.g., AWS, Temi, or Trint) and possibilities for directly coding audio and video files within Qualitative Data Analysis Software (QDAS) packages. Transcribing a focus groups' conversation can be more difficult, as the recordings contain multiple voices. One way to transcribe such data is to "tag" each voice (e.g., Voice A, Voice B). Once the transcription is complete,

DOI: 10.1201/9781032693682-14

researchers can correct any spelling or speech recognition errors, as well as make manual edits, for instance, inserting notations for pauses, laughter, and looks of discomfort and inserting any punctuation. It is also important to anonymise the transcript when possible so that the participants cannot be identified from what was said (e.g., names, places, and significant events).

3. *Becoming familiar with the data* can be considered part of the initial analysis, wherein researchers take note of the ideas or experiences described by participants that appear in interviews, are recorded in observation notes, or are described in documents. These initial understandings often inform later data analysis. Thus, it is important to be aware of and chronicle in detail these early reactions to the data. Becoming familiar with the data also helps a researcher to be aware of the limits or gaps in the collected data. This can inspire future data collection or areas for further research.

4. *Memoing the data* refers to researchers reviewing their data and generating memos that describe initial reflections, as well as any emergent interpretations. Researchers can also note down potential biases that may influence the interpretation of the data. Memos can guide researchers to places in the data where there is, or will be, statements or experiences that are potentially of analytic importance.

5. *Data coding* is the identification of topics, issues, similarities, and differences revealed from the participants' narratives and interpreted by the researcher. Field notes during an interview can be useful complementary information to facilitate this process, as the gap in time between an interview, transcribing, and coding can result in memory bias regarding non-verbal or environmental context that may affect interpretation of data. Other than coding manually, such as with coloured pens to categorise data and subsequently to cut and sort the data, electronic tools (e.g., NVivo) and large language model enabled automatic coding (e.g., GPT-3) are increasingly used for qualitative analysis.

   In the first phase, researchers assign codes to the entire dataset. This layer of coding serves to identify important statements, experiences, and reflections and to reduce the size of the data. In the second phase, researchers frequently return to the passages and data segments to which they assigned codes in the first phase and assign additional codes. Codes at this stage begin moving to a higher level of inference, particularly as a researcher begins to reflect on concepts and/or ideas that are more directly related to the focus of the study. The aim of the second phase of coding is to begin connecting statements, experiences, and reflections offered by research participants to the study's analytic interests. Finally, researchers generally complete a third phase of coding in which they make explicit connections to the conceptual and/or theoretical ideas of the study. In this phase, the coding usually reaches its highest level of inference [Linneberg and Korsgaard (2019)].

6. *Moving from codes to themes*: Metaphorically, codes can be thought of as individual puzzle pieces that, as a collection, contribute to a researcher's depiction of the data. Individually, codes do not tell the entire story. Rather, for that to occur, the researcher must understand how codes (inter)relate and contrast with one another. These (inter)relationships and contrasts are encapsulated in the categories. Categories aggregate individual codes that are related analytically or conceptually. This aggregation represents an important intermediate step in the production of themes. To produce themes, the researcher must bring together various related categories. This involves recognizing the similarities, differences, and relationships across categories. Then, the researcher assigns a statement to these categories. This theme name needs to be inclusive of all the underlying categories, as well as be descriptive of their content, the relationships between them, and any similarities or differences observed. Themes are generally aligned with the conceptual or analytic goals of the study and therefore are designed in response to the study's primary research questions or focus.

12.2 Analysis of Qualitative Data

Qualitative data includes, but is not limited to, open-ended survey questions, semi-structured interviews, and observation notes. The objective of qualitative data analysis is to interpret and structure the meanings and conclusions that can be derived [Thorne (2000)]. Several software tools may support these methods, such as the Provalis Research suite or NVivo. Qualitative data analysis is an iterative process, often performed along with data-collection, where emerging interpretations and observations guide adaptations and expansions of the data-collection process [Busetto et al. (2020)]. Thus, they provide HRI researchers with a suitable means to identify stakeholders' and users' ethical values and requirements, formulate new hypotheses around how users understand and interact with robots, and explore the different contexts (e.g., social, cultural, and personal among others) in which these interactions take place.

There are two main approaches to the analysis of qualitative data. The first approach is to *reduce* the amount or the complexity of the data through identification of core concepts (or codes). Methods such as comparative content analysis, qualitative content analysis, and thematic analysis are suitable for this objective. The second approach aims at *expanding* the data by producing one or more interpretations. Examples of methods that fit into this second group are phenomenological approaches, narrative analysis, and discourse analysis [Flick (2013)].

Although qualitative research encompasses a wide range of methodological strategies and analytical procedures, there are several common data practices shared by most approaches [Lester et al. (2020), Schutt (2018)]:

1. Documenting the data and the process of data collection
2. Assigning codes or themes to the collected data (e.g., field notes, interview transcripts, or documents), as was described in Section 12.2.2
3. Organising and categorising coded materials into concepts, patterns and process, and commonalities and differences
4. Connecting the data to show how concepts influence each other
5. Elaborating a small set of generalisations that cover the consistencies discerned in the data
6. Confronting and documenting generalisations with a formalised body of knowledge as constructs and theories

In the following, we briefly describe the aforementioned methods with examples of their utilisation in HRI research. We refer the reader to Leech and Onwuegbuzie (2007) and Leech and Onwuegbuzie (2008) for reviews of qualitative data analysis methods.

12.2.1 Constant Comparative Analysis

Constant comparative analysis is probably the most commonly used type of analysis for qualitative data [Leech and Onwuegbuzie (2007)]. It aims to generate a grounded theory from a set of themes identified in the collected data using an inductive coding approach (that is, categories and themes are drawn directly from the data). Constant comparative analysis is used to identify patterns in relationships between social actors (e.g., workers and robots) and to identify how interactions between these actors actively construct reality [Suddaby (2006)].

There are three main stages of coding in constant comparative analysis: Open, axial, and selective coding. During open coding, the researcher first reads through the entire set of data, then labels each chunk with a descriptive code, and compares each new chunk of data with previous codes so that similar chunks are labelled with the same code. In the axial coding phase, codes are grouped into similar categories, and connections between categories are established. In the final stage, the researcher waves and refines all major categories into a core category from which a grounded theory can emerge [Kolb (2012)]. Constant comparative analysis is closely related to a grounded theory qualitative research design and is often referred to simply as coding [Leech and Onwuegbuzie (2008)].

In HRI research, constant comparative analysis has been used to generate substantive and exploratory theories on what factors are likely to influence how hospital workers use, perceive, interact with, and accept robots at work [Mutlu and Forlizzi (2008)]. More recently, Meissner et al. (2020) employed constant comparative analysis in a mixture of episodic and problem-centred interviews to identify factors that are likely to influence assembly workers' acceptance of introducing human-robot collaboration at work.

12.2.2 Content Analysis

Content analysis, alongside comparative content analysis, is one of the most commonly used methods to analyse qualitative data. This method is used to systematically compress, describe, and examine trends and patterns in large volumes of qualitative data [Stemler (2000)]. Similarly to constant comparative analysis, content analysis starts by separating the data into chunks and labelling the chunks according to a coding scheme. This type of analysis supports both inductive (i.e., codes emerge from the data) and deductive (i.e., researchers start with preconceived codes or categories derived from existing theories, research, or literature) approaches during coding [Cho and Lee (2014)].

There are two variants of content analysis: Quantitative and qualitative. *Quantitative* content analysis aims at identifying the most frequently used and/or discussed content, as they are strong indicators of the most important concepts found in the data [Leech and Onwuegbuzie (2008)]. This is often done by computing the codes frequency. *Qualitative* content analysis is used to classify large amounts of data into an efficient number of categories that represent similar meanings and identify themes or patterns from these categories [Hsieh and Shannon (2005)]. Although similar to constant comparative analysis, qualitative content analysis is limited in its ability to support theory development. As an example application of content analysis in HRI, Burema (2021) performed a content analysis of 96 academic publications to investigate patterns and trends on how older adults are represented in HRI studies for elder care. Based on this analysis, the author found that the representation of older adults is frequently biased toward presenting older adults in an ageist and neoliberal manner that places most care responsibilities on the individual rather than the welfare state.

12.2.3 Thematic Analysis

Thematic analysis encompasses a wider group of approaches that aim to identify, analyse, organise, describe, and report patterns (themes) within or across qualitative datasets [Lester et al. (2020)]. Themes are constructed meanings and interpretations derived from data that answer a specific research question, rather than summaries or categorisation of codes [Kiger and Varpio (2020)]. A distinguishing characteristic of thematic analysis is its flexibility to be used in a wide range of study questions, designs, and sample sizes. It also allows researchers to discuss points in relation to one another, as opposed to individually, and to provide additional insights about the collected data as a result. Previous studies have suggested that thematic analysis is a more suitable method when working in research teams and analysing large qualitative datasets [Nowell et al. (2017)].

Thematic analysis usually follows the following six-step process [Nowell et al. (2017), Kiger and Varpio (2020), Braun and Clarke (2006)]:

(1) Familiarisation with the data through repeated and active reading of the data,

(2) generation of initial codes,

(3) examination of the codes and data extracts to look for potential themes,

(4) refinement and review of themes,

(5) definition, naming, and narrative description of each theme, and

(6) reporting of the analysis process and results.

Thematic analysis has been identified as a suitable approach to answer questions about the concerns people have about an event or the reasons people use or do not use a service or product [Vaismorady et al. (2013b)]. Most applications of thematic analysis in HRI align with these objectives. For example, thematic analysis has been used to explore and understand the reasons for participants' hesitation when asked to hit a humanoid robot [Riddoch and Cross (2021)], investigate the underlying concerns from primary and secondary users of a human-robot cooperative workstation designed for people with mobility impairments [Arboleda et al. (2020)], and explore the attitudes of older adults and caregivers with regard to the long-term field deployment of socially assistive robots in elder care homes [Carros et al. (2020)].

12.2.4 Phenomenological Approaches

Phenomenological approaches, in contrast to methods aiming to find patterns and commonalities within human experiences, seek to discover the underlying structure and essence of human experiences through the intensive study of individual cases [Thorne (2000)]. Phenomenological approaches are rooted in phenomenological psychology and, as such, seek to understand the subjective meaning of human experience [Knaack (1984)]. That is, they aim to uncover and explicate the deeper human aspects of a situation by looking at a person's lived experience (including mood, sensations, and emotions). To do so, all phenomenological variants (i.e., descriptive phenomenology, hermeneutic phenomenology, and interpretative phenomenological analysis) avoid cross-comparisons and instead target the in-depth and detailed understanding obtained through an exhaustive, systematic, and reflective study of human experiences as they are lived and felt.

There are four key common processes for all phenomenological approaches [Finlay (2014)]:

- *Seeing fresh.* Researchers try to push beyond what they know from experience or through established knowledge so they can remain open to new understandings about the phenomenon of interest.
- *Dwelling.* Researchers get deeply immersed in the minutiae of the data and begin identifying and progressively elaborating a deeper understanding of the participants' experiences.
- *Explicating.* This is the phase of synthesis, integration, clarification, and revelation. Researchers articulate emergent themes into larger themes and/or narratives.
- *Languaging.* Researchers put their findings into words and describe the lifeworld phenomenon as faithfully as possible. The aim is to provide a rigorous, rich description that evokes the phenomenon and the experience of this phenomenon in immediate and potent ways.

In the HRI literature, phenomenological approaches have been used to investigate whether humans can establish romantic relationships with robots [Viik (2020)]. In Parviainen and Pirhonen (2017), drawing on the phenomenology of the body, researchers investigated the embodied interaction in robot technologies for elder care with a focus on older adults with dementia and for whom touch and seeing is a big component of their care. We refer the reader to Finlay (2014) for a detailed introduction to phenomenological analysis and its different variants.

12.2.5 Narrative Analysis and Discourse Analysis

Narrative analysis and discourse analysis aim to study and understand how people present and construct accounts of themselves and their social world through language. Discourse analysis seeks to identify the discourses and interpretative repertoires that individuals use to make sense of their world [Burck (2005)]. In general, discourse analysis involves selecting a representative set of segments of language use (e.g., several lines of an interview), followed by a detailed analysis of these segments to examine the ways in which certain topics are discussed [Leech and Onwuegbuzie (2008)]. Narrative analysis recognises the extent to which stories people tell provide insight into their lived experiences, and thus, through the identification of the main narrative themes within these accounts, it aims to help researchers discover how people understand and make sense of their experiences and lives [Burck (2005)]. From these elements, a narrative (that is, a temporally organised, goal-directed story) is built that details a person's experience in relation to the research question(s) [Sharp et al. (2019)]. An example of discourse analysis applied to HRI studies is Welfare et al. (2019), where the authors explored how workers perceived the positive and negative attributes of their jobs and the deployment of robots within it. The authors devised a set of integration considerations for organisation planning to deploy robotics technology within their manufacturing process.

12.2.6 Choice of Method and Ethical Considerations

The choice of analysis method is determined by multiple factors, such as the type of qualitative data collected, the research question(s) under consideration, and whether or to what extent a chosen method can help answer this research question [Busetto et al. (2020)]. Thus, given the nonlinear, iterative, and often recursive nature of qualitative research, different research designs and analysis methods can be useful within the same study. We provide a high-level comparison of the analysis methods previously described to offer general recommendations for analysing qualitative data in HRI studies.

- *Type of data*: When it comes to the type of data being analysed (e.g., talk, observations, documents), Leech and Onwuegbuzie (2008) provide the following recommendations. When working with conversation data, such as data collected during interviews, narrative analysis and discourse analysis are particularly suitable. Both constant comparison analysis and thematic analysis are highly flexible methods and thus can be used with all types of data. Content analysis is recommended when working with conversation and document data.

- *Researcher experience*: Among the methods listed, thematic analysis has often been suggested as a suitable starting point for researchers who are new to qualitative research. It does not require the theoretical and technological knowledge of other qualitative analysis methods and therefore offers a more accessible form of analysis [Lester et al. (2020)]. Compared to other methods for which a clear set of steps and processes is established, methods such as phenomenological analysis and narrative analysis have limited detailed accounts of the actual techniques and processes undertaken during the analysis. This can make it difficult for novice qualitative researchers to know where and how to start and/or stop when using this kind of approach [Sharp et al. (2019)].

- *Study nature*: Constant comparative analysis is particularly suitable for exploratory research studies, since it allows researchers to look at "phenomena with new eyes and from a new perspective without restriction within already existing hypotheses" [Cho and Lee (2014)]. When working with data obtained from a focus group design, constant comparative analysis, content analysis, and discourse analysis are best suited [Onwuegbuzie et al. (2009)].

- *Research goals*: Analysis methods can be selected according to researchers' goals. For instance, constant comparative analysis can be used whenever the objective is to generate a theory that allows researchers to make sense of people's actions and experiences. If the objective is to describe and summarise the qualitative data collected, as well as to develop a list of categories and themes that describe the meaning of the data, content analysis or thematic analysis may be used [Cho and Lee (2014)]. However, given their descriptive nature, unless they are used in conjunction with existing theoretical frameworks, both content and thematic analyses have limited interpretative power. Discourse analysis is a useful approach to identify the multiple meanings of language, and phenomenological approaches provide a detailed account and descriptions of lived experiences.

Data analysis is often the central step in qualitative research and requires researchers to immerse themselves in data, meaning, and context to understand and provide interpretations of the observed human experiences. Because of this, qualitative data analysis is strongly influenced by theoretical inclinations, epistemological commitments, methodological decisions, personal characteristics, and preconceptions of the researcher [Irwin (2013)]. Furthermore, given the interpretative nature of most qualitative analysis methods, it is critical to guarantee that the participants' views and experiences are not being misrepresented, misshaped, or taken out of context. Several measures can be used to minimise the risk of misinterpretation: Respondent validation (i.e., sharing findings with participants before publication), minimising the researcher isolation during the analysis process by ensuring that novice researchers are adequately trained and supervised or that experienced researchers have sufficient contact with a co-worker who can play the role of "devil's advocate", and explicit disclosure of the researcher's theoretical approach to reduce possible biases [Richards and Schwartz (2002)].

Another aspect to consider during the analysis and reporting of qualitative data is confidentiality and privacy. Qualitative data by their nature are full of clues to participants' identities, so additional care from the researcher is required to ensure the anonymity of participants in published work. Some measures to protect the identity of participants include the use of secure data-storage methods, the removal of identifier components and biographical details amendments, and the use of pseudonyms (applicable to names of individuals, places, and organisations) [Sanjari et al. (2014)].

12.2.7 Statistical Analysis of Qualitative Data

The dominant concept of sample size in qualitative data analysis is saturation. Saturation occurs when adding more participants to the study does not result in additional perspectives or information [Charmaz (2006)]. However, adopting saturation as a generic quality marker may not be appropriate, as qualitative methods often employ purposive sampling that involves information-rich cases. Alternatively, Malterud et al. (2016) proposed the concept of "information power" to guide an adequate sample size for qualitative studies instead of saturation. Information power indicates that the more information the sample has, relevant for the actual study, the fewer participants are needed. They presented a model that considers the purpose of the study, the specificity of the sample, the theoretical background, the quality of dialogue, and the analysis strategy to determine whether sufficient information power will be obtained. A study will need fewer participants when the study objective is narrow, if the combination of participants is highly specific for the study objective, if the study hypothesis is supported by established theory, if the interview dialogue is strong, and if the analysis includes longitudinal in-depth exploration of narratives or discourse details. For HRI studies, we recommend transparent reporting of evaluations of sample size sufficiency, situated within broader and more encompassing assessments of data adequacy. Researchers may define an approximation sample size using previous comparable studies [Mason (2010)]. In addition, the adequacy of the sample size should be continuously evaluated during the research process.

12.3 Mixed-Methods Data and Analysis

12.3.1 Mixed-Methods Analyses

Mixed-methods data analysis consists of qualitative and quantitative analyses under a similar framework [Vogl (2019)], where the degree to which quantitative and qualitative research strands interact with each other depends on the overall design of mixed-methods research [Schoonenboom and Johnson (2017)]. The analysis of one or both data types can occur concurrently in parallel mixed methods, as was discussed in Section 11.2; sequentially in two phases (that is, the qualitative analysis phase precedes the quantitative analysis or vice versa, and findings from the first phase inform the subsequent phase); or in more than two phases (i.e., iterative analysis). It is also possible that the analysis strands will not interact with each other until all results are merged in the interpretation stage [Onwuegbuzie and Hitchcock (2015)].

Mixed-methods analyses can be categorised into standard (non-crossover) and crossover approaches. In a standard mixed-method analysis, qualitative methods are used to analyse qualitative data and quantitative analysis methods are used for quantitative data. In a crossover mixed-method analysis, one form of data (e.g., qualitative) can be analysed using methods traditionally associated with the opposite paradigm (e.g., quantitative), which yields a higher level of integration of both types of analyses than their simple combination during the interpretation stage. When using crossover analysis, it is common to quantify qualitative data or quantify quantitative data. Quantitisation can be done, for example, by computing code frequencies and the percentage of people selecting specific themes, or by looking at whether qualitative responses include certain codes (this is known as dichotomisation) [Driscoll et al. (2007)]. Qualitising involves profiling participants by theoretically grouping or typing them based on their scores from the quantitatively collected data. Qualitative profiling can be done, for example, by attaching a qualitative label to the numerical values of standardised questionnaires or relevant metrics (e.g., respondents with low scores on a five-point Likert-type scale are labelled as in disagreement or participants with low task performance are labelled as novices) [Nzabonimpa (2018)].

In HRI research, it is common to find studies in which both quantitative and qualitative data are analysed in a non-crossover, concurrent, and convergent manner. That is, each data type is analysed independently using methods of the same paradigm (i.e., qualitative data with qualitative methods and quantitative data with quantitative methods), and results are later compared or combined with the aim of gaining a more complete understanding of a problem and/or validating one set of findings with the other. For example, Björling et al. (2020) analysed questionnaire data and interviews to understand the effect of three different robot embodiments (physical, computer screen, and virtual reality) on adolescents' momentary stress during a simple, stress-disclosure activity. In this study, quantitative analysis of the results of the questionnaire offers an overview comparing the effect of the different robot embodiments, while qualitative analysis of the interviews provides a detailed explanation of the observed differences. Some examples of this convergent analysis can be found in Björling et al. (2020) and Pedersen et al. (2020).

Crossover mixed-methods analysis can be further decomposed into quantitative-dominant, qualitative-dominant, or equal status. While quantitative- or qualitative-dominant analyses are performed under the assumption that the inclusion of data and analysis methods from the alternative paradigm can help address the research question(s), in equal-status crossover mixed-methods analysis, both types of analysis are used to optimally address the underlying research question(s) [Onwuegbuzie and Hitchcock (2015)]. We refer the reader to Onwuegbuzie and Johnson (2021) for a detailed review of quantitative-dominant, qualitative-dominant, and equal-status methods.

12.3.2 Choice of Method and Ethical Considerations

Mixed-methods analyses expand and strengthen the observations and conclusions that researchers can obtain. However, there are complexities associated with their use. For instance, they require expertise in both qualitative and quantitative data analyses, as well as additional intellectual, financial, and time resources. In addition, researchers face additional ethical challenges in data integration, method selection, and contradictory results between analysis strands. For example, the integration of qualitative and quantitative data can increase the chance of participants being identified and linked to their data. Researchers are also at risk of privileging one method over another [Preissle et al. (2015)].

Mixed-methods analyses require researchers to make informed and transparent decisions on combining qualitative and quantitative analyses. Vogl (2019) recommended that researchers consider two things when deciding which mixed-methods analysis to use: The purpose of the analysis and how to implement the analysis. Regarding the first consideration, the purpose of mixed-methods analyses can be described based on Schoonenboom and Johnson (2017), namely, credibility (i.e., the use of both analysis strands should enhance the integrity of the findings), context (i.e., the combination of analysis methods provides contextual understanding), illustration (i.e., qualitative data helps to illustrate quantitative findings), utility (i.e., combining both approaches will be more useful to practitioners and others), confirm and discover (i.e., qualitative analysis can generate hypotheses and quantitative analysis can test these hypotheses within a single study), and diversity of views (i.e., researchers' and participants' perspectives are combined through quantitative and qualitative research, respectively, or relationships between variables are uncovered through quantitative research while meanings among participants are revealed through qualitative methods).

With regard to the implementation of a mixed-methods analysis, there are four key factors to be considered: The degree of interaction between the analysis strands (e.g., standard or crossover analysis), the relative weight or importance attributed to each strand (e.g., quantitative-dominant, qualitative-dominant, or equal status), the temporal relation of the strands (e.g., sequential or concurrent), and how to combine or integrate them (e.g., integration in the interpretation phase or during analysis) [Vogl (2019)]. Both choices must be guided by the research question(s) and the study design.

We end this section with a list of software tools that can be used to perform mixed-methods analyses. These tools are as follows. Provalis Research ProSuite[1] includes three software tools: QDA Miner for qualitative data coding and analysis, WordStat for content analysis and text mining, and SimStat for general statistical analysis. MAXQDA[2] is a software package for both qualitative and mixed data, and it offers the option of analysing each quantitative and qualitative data separately or in an integrated manner. ATLAS.ti[3] is a multiplatform tool for the analysis of large bodies of qualitative data (i.e., text, graphics, audio, and video). Finally, NVivo[4] also supports qualitative and mixed-methods data and, in addition, supports logic-based queries that help researchers to identify connections in their data.

12.4 Summary

We summarised common practices in analysing qualitative data, namely, documenting the data and the process of data collection, transcribing and familiarising oneself with the data, and assigning codes or themes to the collected data (e.g., field notes, interview transcripts, or documents). We reviewed commonly used analysis methods and provided general guidance on the questions and aspects to keep in mind when choosing and/or using a specific analysis method. Our review aims to introduce these qualitative and mixed-methods data analysis to the HRI field to help researchers understand the users of robots and the context of human-robot interactions.

Recommendations for Data Collection and Analysis of HRI Studies

- Qualitative data analysis is a non-linear, iterative, and recursive process. When deciding which method to use, researchers must consider the type of qualitative data, their experience using qualitative analysis methods, the study nature, and the research goal. An effective way to help with this decision process is to build collaboration with an interdisciplinary team. This will also help to ensure the quality of the qualitative data and analysis collected.
- Mixed-methods analysis helps to expand and strengthen the observations and conclusions that researchers can obtain from their studies. However, this type of analysis requires more resources than purely qualitative or quantitative studies.
- Independently of the analysis method being used, researchers are encouraged to open-source all information necessary to reproduce their analysis and results.

Notes

1. https://provalisresearch.com/
2. https://www.maxqda.com/
3. https://atlasti.com/
4. https://www.qsrinternational.com/nvivo-qualitative-data-analysis-software/home

Part III

Case Studies

13

Participatory Design with Stakeholders

This chapter discusses how participatory design is used in HRI studies to understand the functional and ethical requirements of stakeholders. We examine an example HRI research project where a series of participatory design workshops were conducted to investigate expectation and perception of the general public towards social robots deployed in public spaces. This study demonstrates how quantitative and qualitative methods can be used to understand users' emerging perceptions of a robot's capabilities through contextualised design-evaluation cycles.

13.1 Participatory Design

As discussed in Section 11.1.7, participatory design (PD) is a method for exploration and incorporation of stakeholders' ethical and functional requirements. In PD, participants are included in early stages of the research and design process, instead of later in the evaluation stage.

As detailed in Table 11.1, PD often involves three phases: initial exploration focusing on understanding the context, discovery processes focusing on understanding the users, and prototyping focusing on iterative design and generative activities. Common data-collection methods in PD include interviews (Section 11.1.2), focus groups (Section 11.1.3), questionnaires (Section 5.2.5), and observations (Section 11.1.1). The success of a PD study relies on the researcher's ability to facilitate productive collaboration with the participants and extract informative insights from the PD activities.

DOI: 10.1201/9781032693682-16

13.2 Case Study: Robots in Public Spaces

As an example of adopting PD methodology for HRI research, we discuss a series of studies where PD was employed to investigate expectations and perceptions of the general public towards robots deployed in public spaces, such as a shopping mall [the Robots in Public Spaces (RiPS) project[1]].

13.2.1 Identifying Stakeholders and Research Questions

Consistent with existing guidelines [Axelsson et al. (2021), Spinuzzi (2005)], the RiPS project was conducted by a multidisciplinary researcher team from the fields of robotics, design, human geography, and policy development to facilitate successful PD. As a standard practice (Section 4.6.3), the research protocols were reviewed and approved by the researchers' affiliated university's human research ethics committee before the experiments, with each iteration of research methodology revision reviewed and approved via ethics approval amendments.

As was discussed in Section 2.3, the research team identified that people increasingly encounter robots deployed in public spaces. However, they can rarely provide their inputs when the robots and their tasks and behaviours are being designed. This has motivated the need to study what the general public expects of such robots, as well as how their experience of public spaces may be influenced by the presence of the robots and their direct or indirect interaction experiences with the technology.

A diverse participant population is necessary to capture the general public's ethical and functional requirements of robots in public spaces. Due to feasibility considerations, the participants recruited for the RiPS project were limited to the general population at the researchers' affiliated university. To increase representation, participants from all faculties were invited, particularly those without prior experience with robots or programming. Participants were recruited through word of mouth from the personal and social networks of the research team and previous research participants. For the first and second PD workshops, 12 participants (7 females, 5 males) from the architecture, arts, criminology, and other faculties attended both workshops. For the third workshop series consisting of eight individual sessions, an additional 21 participants (11 females, 10 males) from various faculties were recruited. All participants provided their written consent for participation and data collection after being fully informed on the research aim and study protocols.

13.2.2 Study Aims and Methodological Motivations

As was discussed in Section 11.1.4, in existing HRI research adopting PD with generative activities of developing robot prototypes, participants often engage in paper-based methods with open-ended objects, for example, sketches of envisioned interaction scenarios and robot designs using figurines. However, such an approach may lack grounding to the actual capabilities achievable by existing robotic technologies. It has been found that participants without technical backgrounds often overestimate a robot's capabilities due to influences from science fiction and the media [Horstmann and Krämer (2019)]. Thus, it can be difficult to translate insights gathered from such PD sessions to currently available robots. Furthermore, a tangible realisation of the designed prototype is crucial for understanding any requirement or perception changes a participant may experience after interacting with a redesigned robot. Therefore, the goal of PD in the RiPS project was to enable participants to prototype robot behaviours while grounding their experience with the available technology. In the following sections, we describe the evolution of the PD methodology in the RiPS project, as well as the data analysis approaches.

13.2.3 Evolution of the PD Methodology in the RiPS Project

The majority of the RiPS project was conducted in 2020–2021 amidst the COVID global pandemic. Due to strict lockdown measures in Melbourne where the research team and participants were located, the PD activities were conducted remotely as online workshops using the video-conferencing platform. The design of the PD workshop procedure went through two iterations, starting with exploratory discussion and generative activities using a robot simulator [Tian et al. (2020)], with a follow-up generative activities session using a physical robot live-streamed to the participants [Carreno-Medrano et al. (2021), Tian et al. (2021)], and finally expanding to a series of PD workshops combining the simulated and physical robot using two different robotic platforms with smaller focus groups in each session [Carreno-Medrano et al. (2024)]. In addition to these PD workshops with focus groups, one-on-one interviews [Sumartojo et al. (2021)] were conducted to gather individual reflections.

13.2.3.1 First PD Workshop with a Simulated Robot

In the first PD workshop (WS1) [Tian et al. (2020)], a three-phased structure was adopted, including an open discussion, a participatory prototyping session using a simulated robot, and follow-up one-on-one interviews, as shown in Figure 13.1.

The 30-minute open discussion focused on the initial exploration and discovery processes of PD. Its goal was to collect initial expectations of the participants towards robots in public spaces. Before attending the workshop,

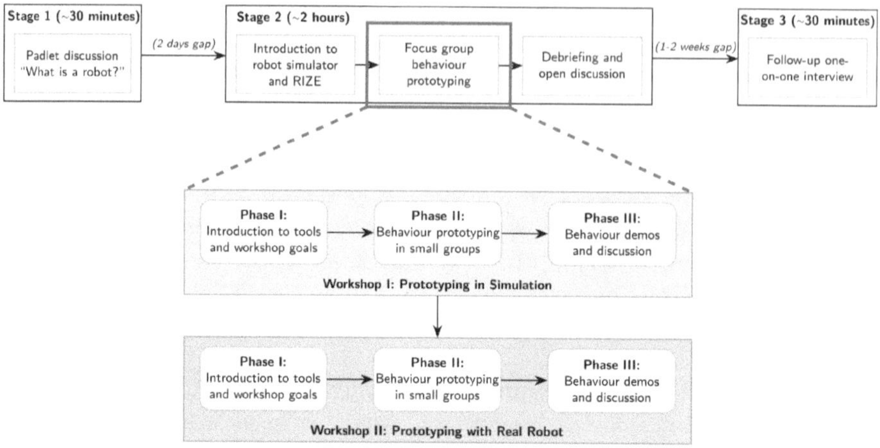

FIGURE 13.1
Structure of the first and second PD workshops.

participants were asked to each upload at least one image that captured their understanding of "a robot in public space" to the Padlet platform online, as shown in Figure 13.2, which became the basis of the in-workshop focus group discussion. As was discussed in Section 11.1.3, in this focus group discussion, one researcher took the role of a facilitator to lead the discussion with a set of open questions, while others from the research team performed logistic tasks and contributed to asking follow-up questions.

FIGURE 13.2
Open discussion phase of the first online PD workshop. [From Tian et al. (2020). Reprinted with permission from Springer Nature.]

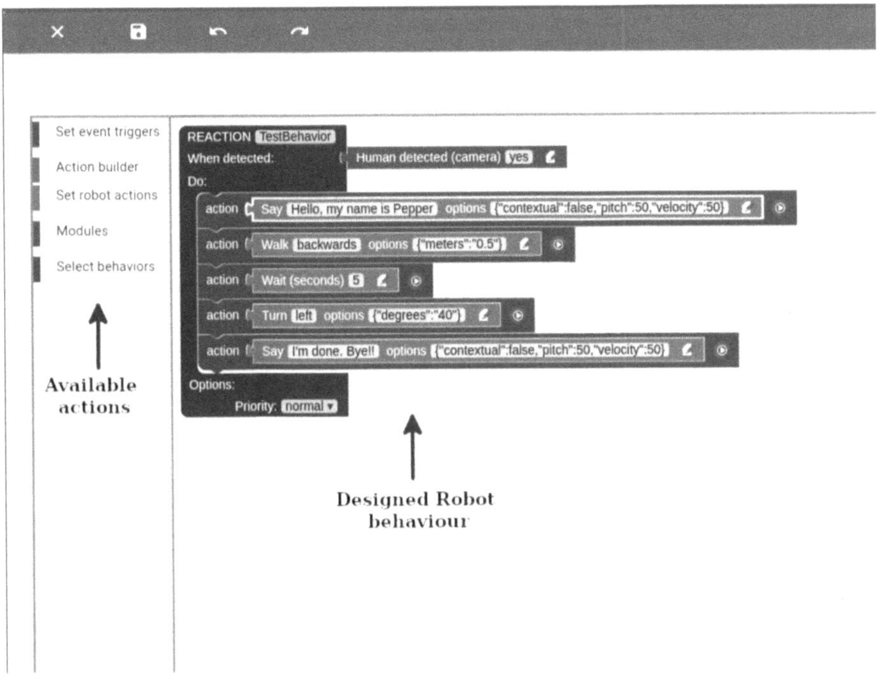

FIGURE 13.3
RIZE end-user programming interface. [From Carreno-Medrano et al. (2021).]

Having each participant submit at least one image ensured a balanced discussion for the relatively large focus group size ($N = 12$).

The 2-hour prototyping session was conducted 2 days after the open discussion and focused on eliciting user-centred discussion with iterative robot behaviour design and revision. This phase can be further divided into three steps, namely, an initial orientation and introduction to the prototyping resources, hands-on behaviour prototyping, and debriefing and demonstration for general discussion. In the generative activities of robot behaviour prototyping (see Section 11.1.4), the RIZE end-user programming interface [Coronado et al. (2021)], illustrated in Figure 13.3, was employed to support participants who did not have technical knowledge on robot programming. This allowed participants to easily understand and design complex robot behaviours by combining robot action primitives as building blocks.

In addition, a simulated Pepper humanoid robot, as shown in Figure 13.4, was used for participants to visualise the execution of their designed robot behaviours. The choice of the robot platform was based on both available resources and as a representative robot currently adopted for the public space context. The robot simulator allowed exploration of available robot functions and revisions of the robot behaviour prototype. During the prototyping step,

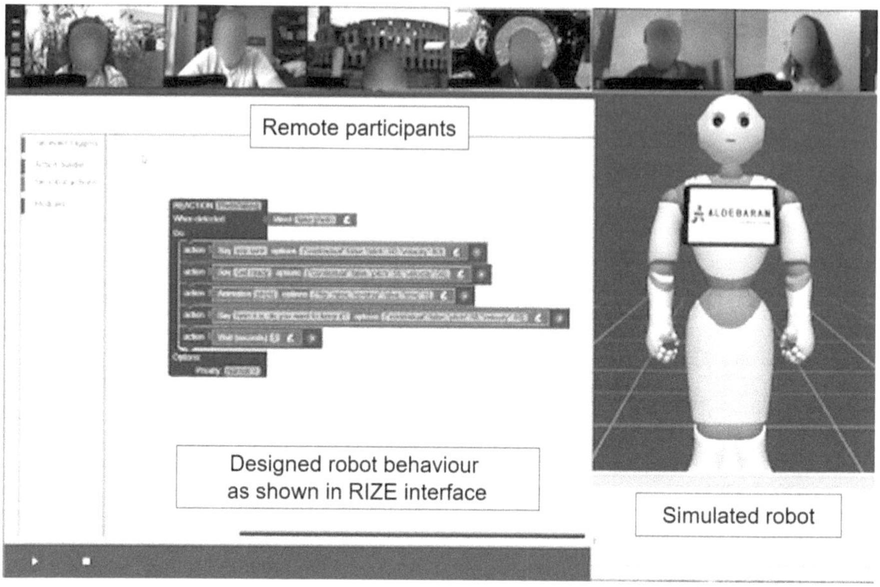

FIGURE 13.4
Participatory prototyping with a simulated robot in the first online PD workshop.

the participants were divided into two smaller focus groups ($N = 6$) using the breakout room function of Zoom to allow effective prototyping, with each focus group supported by a facilitator and a technical expert from the research team. Separating the facilitator and technical expert roles allowed the facilitator with experience in interviews and PD to focus on guiding the participants' discussion and extracting insights, while the technical expert with experience in robot programming could focus on providing information related to the robot's capability and supporting the prototyping activities. Thus, this design facilitated a productive PD.

In the third phase, follow-up individual interviews were conducted within 2 weeks after the prototyping session to collect personal reflections of the participants. These 30-minute interviews followed the interview guide approach (see Section 11.1.2): A researcher followed a list of topics of interest to guide the participant to present his or her reflection on the PD activities design and perception towards the robot and the interaction context.

13.2.3.2 Second PD Workshop with a Physical Robot

Based on insights drawn from the first PD workshop, a follow-up study (see Chapter 7) was designed. The same 12 participants from WS1 attended the second workshop (WS2). It expanded the prototyping session (second phase

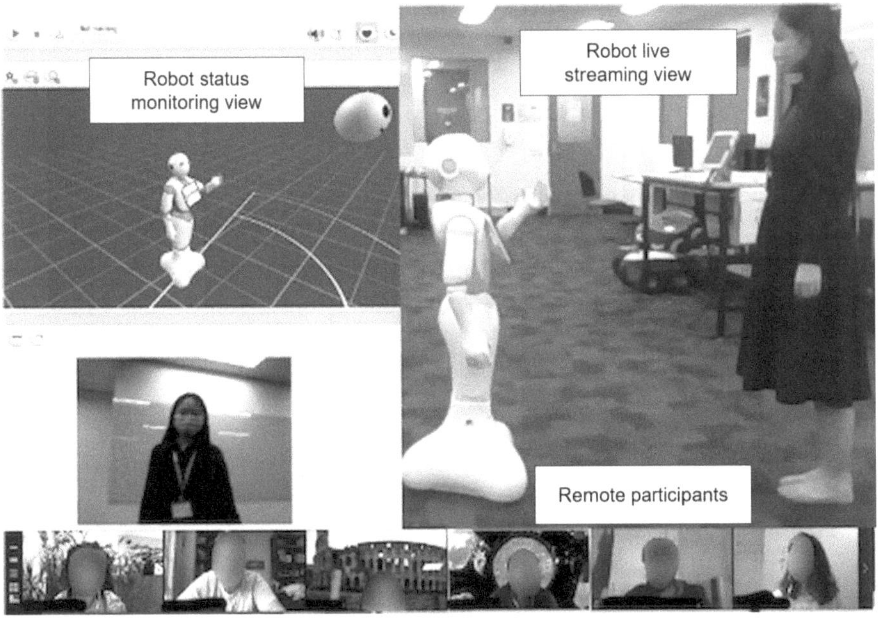

FIGURE 13.5
Prototyping with a physical robot in the second online PD workshop.

of WS1, as shown in Figure 13.1) from the simulated robot to a physical robot [Carreno-Medrano et al. (2021)], as demonstrated in Figure 13.5. This extension served two main objectives: (1) It allowed participants to observe the robot's behaviours within the social-physical context, and (2) it supported the design of interactive robot behaviours not achievable in a simulated environment.

Similar to WS1, during the prototyping step, the participants were divided into two focus groups. A cheat sheet listing all available robot actions (see Figure 13.6) was provided as an additional programming resource. Further, the technical support researcher in WS1 was formalised into a *translator* role, who assisted with the implementation and answered participants' questions regarding the programming interface or the robot's capabilities.

In addition to the focus group discussion, a subjective questionnaire (see Section 5.2.5) was added before and after WS2 to collect quantitative data on participants' perceptions towards the robot, namely, the GodSpeed questionnaire [Bartneck et al. (2009)] for likeability and perceived intelligence, the Trust questionnaire [Muir (1994)], the Reliability questionnaire [Sidner et al. (2005)], and an ad hoc questionnaire evaluating how successful participants expected the robot to be at executing a designed behaviour and achieving its intended outcomes.

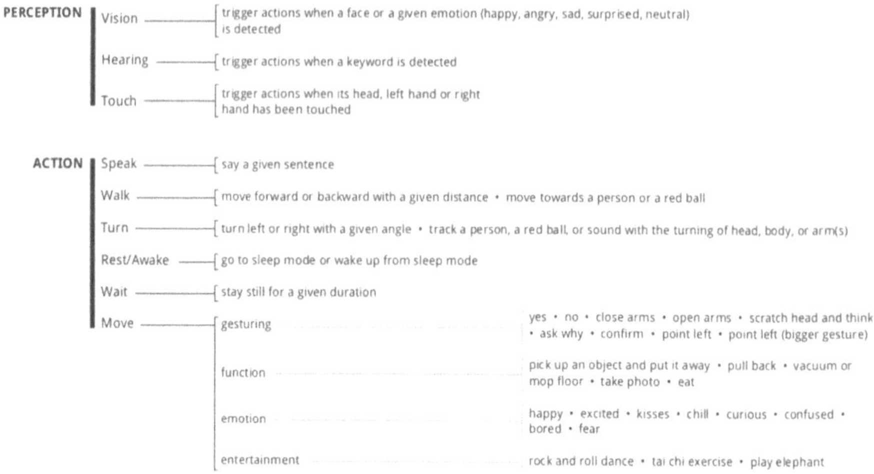

FIGURE 13.6
The robot action cheat sheet for supporting participatory prototyping.

13.2.3.3 Workshop 3 Series with Two Robots and Mini Focus Groups

Building on WS1 and WS2, the PD methodology was further developed in the workshop 3 series (WS3) to combine the benefits of the rapid design-evaluation cycle enabled by the robot simulator together with the incorporation of physical and interactive contexts facilitated by the physical robot [Carreno-Medrano et al. (2024)], as shown in Figure 13.7. Each WS3 session lasted 2 hours. To retain the exploration and discovery process in the compact WS3 format and contextualise the PD activities, in the

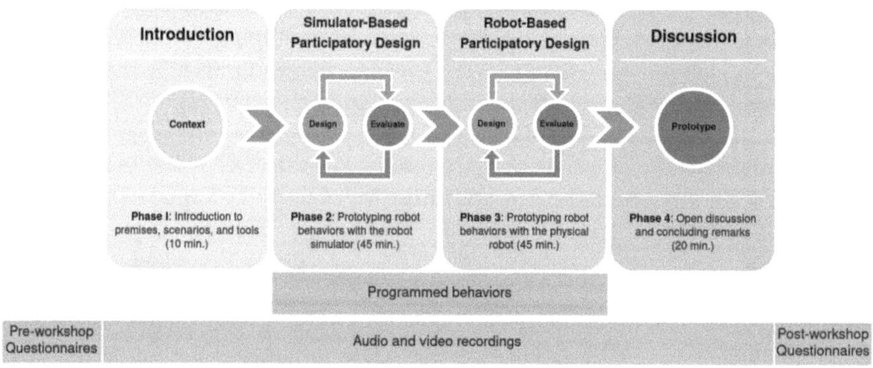

FIGURE 13.7
Structure of an individual PD workshop in the workshop 3 series.

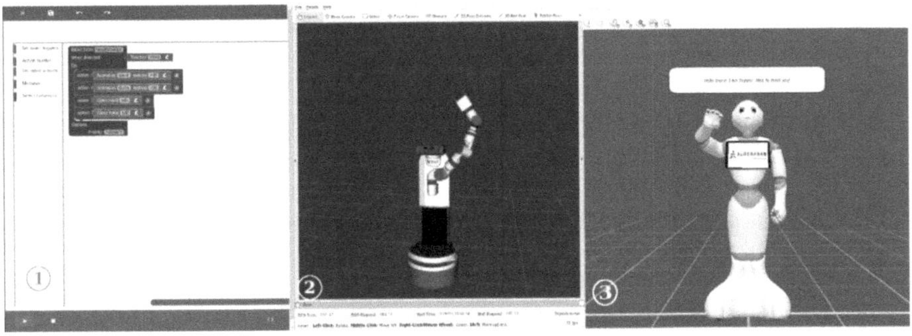

FIGURE 13.8
The (1) RIZE interface, (2), Fetch robot, and (3) Pepper (3) robot used in WS3 (simulator view).

introduction step of WS3, photos of public spaces on campus that were familiar to the participants were used.

In addition to the Pepper humanoid robot used in WS1 and WS2, WS3 introduced the Fetch mobile manipulator robot (see Figure 13.8) to allow investigation on how a robot's appearance may influence participants' designs and perception. Further, the size of the focus groups was reduced to two to four participants (i.e., mini focus groups) to allow for increased contributions and engagement from individual participants.

A total of eight focus group sessions were conducted in WS3, with 21 participants recruited from the university's staff and student population: 10 participants joined the four sessions using the Pepper robot (WS3-Pepper-1 2f0m, WS3-Pepper-2 0f3m, WS3-Pepper-3 1f1m, and WS3-Pepper-4 2f1m), and 11 participants joined the four sessions using the Fetch robot (WS3-Fetch-1 1f1m, WS3-Fetch-2 2f1m, WS3-Fetch-3 and 0f2m, and WS3-Fetch-4 3f1m).

Compared to WS1 with a focus on qualitative analysis, conducting multiple PD sessions in WS3 facilitated quantitative analysis in addition to the pre- and post-workshop questionnaires (similar to those used in WS2). Specifically, the RIZE end-user programming interface was revised to allow for detailed logging of robot behaviour primitives added, removed, or revised during prototyping with precise timestamps. Audiovisual recordings of the workshops were transcribed and annotated to allow quantitative analysis of the discussion with off-the-shelf natural language processing tools.

13.2.3.4 Speculative Interviews

Recall that WS2 and WS3 focused on the generative activities of robot behaviour prototyping. To continue the exploration and discovery processes of PD, which were conducted in the WS1 Padlet open discussion and one-on-one follow-up interviews, parallel to WS2 and WS3, individual speculative interviews were conducted [Sumartojo et al. (2021)]. These interviews focused on policy and social-geographical implications of public space robots, as opposed to technical and user perception considerations of robot design.

In particular, 18 participants (10 female, 8 male) recruited via the researchers' social and professional networks were interviewed online using Zoom in 30-minute one-on-one interviews. These interviews employed the imaginative approach, in which modified photos were used as "visions of the future" to understand people's view on emerging technologies. The imaginative approach is similar to how simulation and Wizard-of-Oz approaches (see Section 3.2.4) can be used to circumvent existing technical limitations in developing and evaluating robot functions.

The images were created using collage, an art creation process that assembles a collection of elements. As shown in Figure 13.9, the images demonstrate different robots and people in a diverse selection of outdoor and indoor environments. The background of each image was created from a photograph representing a public space in the home city of the participants, which they were familiar with and may attend routinely for various activities. Robot cutouts from publicly available digital images and architectural cutouts of human figures were then layered onto each background. The scenarios depicted were designed to represent possible direct or indirect HRI based on existing operations of the selected robots at similar sites for various purposes. Finally, an oilifying filter was applied to allow the images to be non-specific, thus open and ambiguous to evoke conversation.

The participants were interviewed on their overall impressions of the images, what they thought the robots were doing, how they felt about these activities, and whether the robots "felt right" in the contexts of the images. Each participant was asked about all the images. The interviews were analysed with thematic analysis on three aspects: other technologies with which participants were already familiar; the spatial context depicted in the scenarios and their lived experiences of those contexts; and a sense of their own bodies and data, particularly their vulnerability and resilience.

13.2.4 Data Analysis Approaches in the RiPS Project

As was described in Section 13.2.3, the RiPS project employed both qualitative and quantitative methods for analysing the PD data. In particular, WS1 [Tian et al. (2020)], WS2 [Carreno-Medrano et al. (2021), Tian et al. (2021)], and the additional speculative interviews [Sumartojo et al. (2021)] focused on qualitative analyses of the focus group discussion and individual interviews, while the WS3 series included quantitative analyses of the prototyping activities and linguistic analysis of the discussion transcripts [Carreno-Medrano et al. (2024)]. Further, findings of the PD workshops and interviews were reviewed from the policy [Mintrom et al. (2022)] and human geography perspectives [Sumartojo et al. (2023)] in addition to the context of HRI.

FIGURE 13.9
Collages of robots in outdoor and indoor public spaces as speculative interview stimuli. [From Sumartojo et al. (2021). Reprinted with permission from Elsevier.]

13.2.4.1 Qualitative Analysis of PD

In WS1 [Tian et al. (2020)], thematic analysis (see Section 12.2.3) was performed to identify the main themes emerging from the focus group discussion and prototyping activities. In the open discussion stage, researchers reviewed the participants' discussion and submitted captions of the Padlet image submissions. Rooted in the research question of understanding people's expectation and perception towards robots in public spaces, two main themes were identified: (1) A participant's expectation towards robots is shaped by personal experience and popular culture, and (2) participants reported anxiety and mistrust around robots in public spaces.

The observations of participants' prototyping activities as well as the two groups' discussions during stage 2 of WS1 were reviewed by the researchers facilitating each group. The key insights were then discussed within the research team. This resulted in five main themes uncovered, including participants' choice of context and robot behaviour design, shared robot behaviours designed by the two groups, influence of societal context on the robot behaviour design and participants' perception towards it, grounding of expectations via hands-on prototyping, and reported limitations of the workshop design.

Thematic analysis was employed by the interviewer to identify key topics emerging in the follow-up interviews of WS1 and the speculative interviews conducted in parallel to WS3. For WS1 interviews [Tian et al. (2020)], as the aim was to gather participants' personal experience and reflection on the PD workshop, the main themes centred around participants' expectations towards the PD activities and the robot, their report on the limits and benefits of technical tools enabling the prototyping, and the appropriateness of the selected robotic platform for the envisioned public space context and functional purposes.

For the speculative interviews using images of envisioned robot deployment scenarios [Sumartojo et al. (2021)], the research question shifted to understanding people's acceptance of robots in public spaces. Thus, the main themes identified changed according to how participants draw from familiar technology to understand public space robots, the impact of the spatial context, and concerns around potential risks of the technology interacting with vulnerable populations and/or sensitive data.

13.2.4.2 Quantitative Analysis of PD

In WS2, beyond thematic analysis [Tian et al. (2021)] as adopted in WS1 and the one-on-one interviews, inductive coding, a quantitative approach to content analysis (see Section 12.2.2), was performed on the workshop recording by the two translators [Carreno-Medrano et al. (2021)]. To compute inter-coder reliability, approximately 60% of the data was annotated by both translators independently and the kappa coefficient

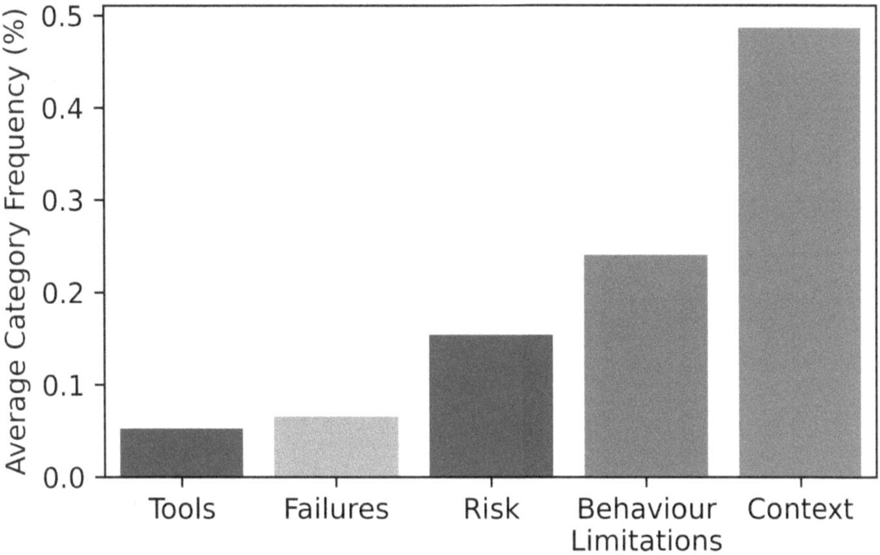

FIGURE 13.10
Distribution of inductive codes associated with each theme [From Carreno-Medrano et al. (2021).]

between their annotations was computed. The result indicated a substantial agreement between the two annotators ($k = 0.64$, 95% confidence interval [0.59, 0.70]).

The annotated codes were aggregated into five higher-level themes of the workshop based on the consensus of the annotators, namely, context complexity, behaviour limitations, risk concerns, performance and interaction failures, and tools limitations. By counting the occurrences of the codes associated with each theme in the annotated workshop transcripts, the frequency of each theme being mentioned by the participants was visualised, as shown in Figure 13.10. The results highlighted that the social and spatial contexts were a key factor in people's perceptions towards public space robots, as well as people's desire for the robot to have more capabilities than it current has.

Recall that in WS2, participants filled in a questionnaire before and after the PD workshop. Thus, by comparing the distribution of their ratings, additional insights can be drawn on how PD influenced their perception towards the robot. Because some of the questionnaire data violated the normality assumption required for a parametric t-test, non-parametric procedures were used in the statistical analysis (see Section 10.2.1). Similarly, after validating that the data were non-symmetric, a two-sided paired sign-test was chosen. This test determines whether the median difference between the participants' ratings before and after WS2 is not equal to 0 (alternative hypothesis).

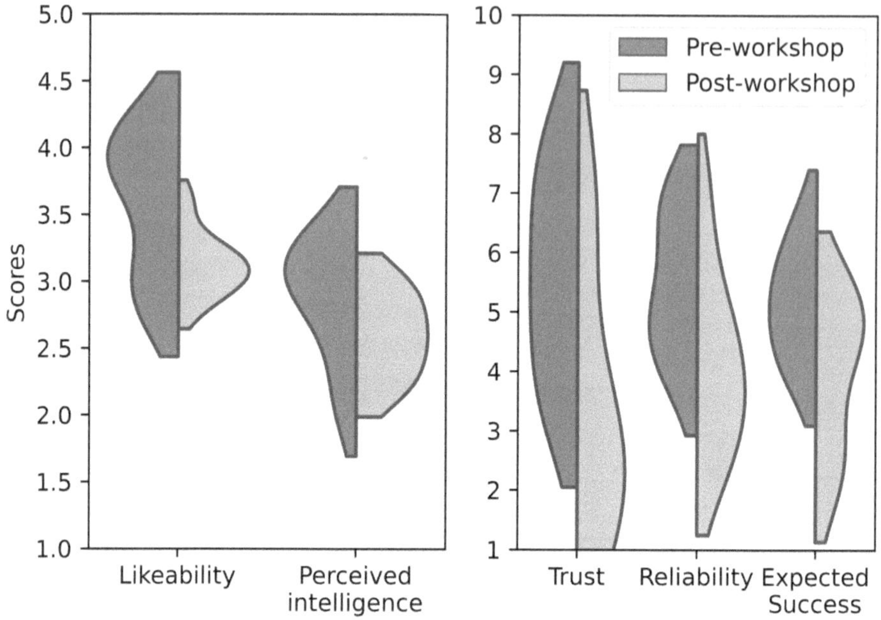

FIGURE 13.11
Distribution of pre- and post-workshop questionnaire ratings. [From Carreno-Medrano et al. (2021)].

A trend of decreasing ratings was identified from visualisation of the data distribution, as shown in Figure 13.11. However, the paired sign-test only found statistically significant rating decreases ($p < 0.05$) for Perceived Trust, Reliability, and Expected Success ratings (Likeability and Perceived Intelligence had $p > 0.05$). This may be a result of the small sample size of the questionnaire responses (see Section 10.3) in WS1 and WS2 ($N = 12$).

In WS3 [Carreno-Medrano et al. (2024)], the quantitative analysis is further facilitated by timestamped prototyping action logs and data-driven analysis of the workshop transcripts, in addition to WS2's methods based on questionnaire ratings and inductive coding. The RIZE end-user programming interface was used to track the types of robot behaviour primitives selected by the participants (see Figure 13.12 for an example), as well as insertions, deletions, and edits made to the behaviour tree representing the prototype robot application (see Figure 13.13). The prototype behaviours developed in the robot simulator and on the physical robot at the end of each design-evaluation iteration were then compared across the groups to investigate how participants' preferences evolved as a result of their emerging understanding of the robot's capabilities for the application scenarios they envisioned.

The word frequency and sentiment of the WS3 transcripts were analysed with off-the-shelf natural language processing tools to gather linguistic

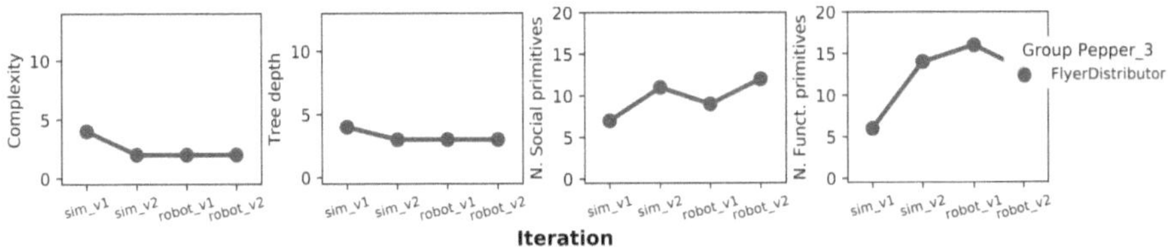

FIGURE 13.12
Evolution of the robot behaviour tree designed by WS3 Pepper Group 3, in which the robot was designed to distribute flyers at a public space.

insights on the focus group discussion. Further, causal relations between thematic codes annotated via a manual inductive coding process similar to that in WS2 were computed to allow a data-driven approach to discovering the inter-dependencies between participants' perceptions towards the robot and their design decisions. These quantitative approaches facilitated behavioural measurement (see Section 5.2.2) of PD in addition to the qualitative approaches.

13.2.4.3 Reporting Insights for HRI and Beyond

Findings of the RiPS project were reported mainly to the HRI community regarding emerging user perceptions contextualised by interaction with robots. To support reproducibility (see Chapter 7), the RIZE interface was open-sourced with documentation to support installation and usage by other researchers. Supporting materials, including the questionnaire, the robot programming cheat sheet, and scripts for quantitative analysis of the transcripts, were included as supplementary materials to the description of the PD method in the publications. However, due to privacy considerations, raw recordings and transcripts of the focus groups, which contained identifiable data of the participants, were not shared beyond the research team.

Beyond the HRI community, insights from the PD workshops were reviewed by members of the research team with a social science background. The findings of the RiPS project that focused on participants' discussion of

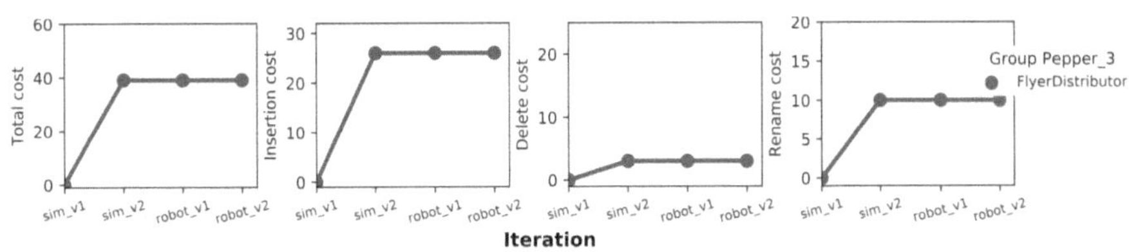

FIGURE 13.13
Revisions to the robot behaviour tree by WS3 Pepper Group 3.

how robots influence the social and spatial dynamics of a public space were reported to the human geography discipline [Sumartojo et al. (2023)]. Based on participants' discussion of concerns and risks towards robots, a policy design checklist [Mintrom et al. (2022)] was proposed to capture key issues related to ethical and effective public space deployment of robots, together with suggested policy responses to such issues, as shown in Table 13.1.

TABLE 13.1
Key Issues of Robots in Public Spaces with a Matching Policy Checklist Questions and Suggested Responses

Key Issues	Checklist Questions	Policy Responses
Safety	How to ensure safety?	Regulation, monitoring, enforcement
Privacy and ethics	How to address privacy and ethics concerns?	Regulation, monitoring, enforcement
Productivity	How to advance productivity?	Regulation, provision of incentives
Aesthetics	How to attend to aesthetics?	Monitoring, co-design
Co-creation	How to encourage co-creation?	Provision of incentives
Equitable access	How to promote equitable access?	Monitoring, subsidies
System innovation	How to facilitate systemic innovation?	Monitoring, funding

13.3 Discussion

Here we discuss the strengths and limitations of the RiPS studies.

13.3.1 Strengths

By employing the PD methodology, the RiPS research team and end-users engaged in a productive mutual communication on designing and evaluating public space robots. This reflective process also enabled continuous improvement of the PD workshop design. Compared to existing HRI studies with PD that remained at the conceptualisation and exploration stage, the active and iterative prototyping session of the RiPS PD sessions provided the participants with a tangible demonstration of their design implemented on an existing robot, thus grounding their perception and expectation with current robots' capabilities. The multidisciplinary research team of the RiPS project allowed for effective development and facilitation of the PD sessions, benefiting from the technical expertise of robotic researchers as well as the moderation and interview skills of social science researchers. This joint effort also resulted in the expanded data analysis approaches utilising both quantitative and qualitative methods to investigate the complex research questions. The supplementary materials and open-source programming tools increased the transparency and reproducibility of the studies, allowing other HRI researchers to build on top of the PD methodology developed in the RiPS project.

13.3.2 Limitations

A major weakness of the RiPS studies is the limited representativeness of participants recruited for the PD workshops. Because the participants were recruited mainly within the university population, even though the majority of them did not have programming or robotics experience, most participants had a higher education background with interests in robotics technologies (see Section 4.2 for more discussion on participant sampling). Thus, they may have expressed more optimistic expectations towards robots or reported more positive experiences during PD. Moreover, as PD draws on a participant's personal experiences, the limited diversity in the participants may have constrained the discussion on important ethical and functional requirements. For instance, the theme on accessibility of robots and other technologies in public spaces was discussed in a limited capacity, such as in terms of potential issues of a robot misrecognising the emotional states of a user from less represented or vulnerable groups [Tian et al. (2021)]. Further, the participants recruited represented end-users engaging in intentional or incidental interaction with public space robots, while additional stakeholders, including staff in the public space, service providers that deploy the robot, and robot developers, were not included. For example, in some prototype behaviours, the participants designed the robot to contact human staff to transfer complex or sensitive service requests [Tian et al. (2021)]. However, the PD session was unable to further investigate how the robot might collaborate with a human worker and/or draw their attention to the request from the perspective of this stakeholder group. In Chapter 14, we will use a case study to further discuss recruitment of domain experts and a specific user population.

Another limitation of the focus group approach to PD is that the discussion is influenced by group dynamics. While having an experienced facilitator helps direct and balance the discussion, the prototyping activities can still gravitate towards the first proposal or the design decisions of those more actively engaged in the discussion. Further, the focus group setting is less suited for discussing sensitive or personal topics compared to individual interviews, as was discussed in Section 11.1.3. In addition, both focus group discussions and interviews are influenced by the interests and background of the facilitator of interviewer. Similarly, thematic analysis of the PD sessions and interviews is subject to the interpretation of the annotator, that is, the researcher bias, as was discussed in Section 10.1. However, the combination of various study methods and data analysis approaches in the RiPS studies helped reduce this limitation to some extent.

Finally, due to constraints in resources and feasibility, the prototyping in the RiPS studies was limited to lab-based demonstration and evaluation of the robot behaviours within the time frame of the 2-hour PD workshop session. The design-evaluation iterative cycle is expected to continue if a designed robot behaviour is deployed in the envisioned application scenario with a specific social and spatial context and when the robot is deployed for an extended period. Such in-the-wild, long-term evaluation is especially important for understanding contextual and longitudinal outcomes of HRI, as was discussed in Section 3.2.2. As PD sessions are costly to conduct, a key challenge is to develop new PD designs that allow for the scaling up of the methodology both in terms of participant population and experiment frequency and duration.

13.4 Summary

In this chapter, we reviewed the Robots in Public Spaces (RiPS) project as an example of using participatory design (PD) in HRI. A series of focus group discussions, participatory prototyping sessions, and one-on-one interviews were conducted to investigate the research questions of what do end-users expect of robots deployed in public spaces, how they perceive such robots' roles and purposes, and how to design robot behaviours to be functional as well as acceptable in the deployment context. We summarised the evolution of the PD workshop methodology expanding from simulator-based to using a simulator and a physical robot, from a single robot to two different robot designs, and from a single focus group to a set of compact mini focus groups. We also reviewed the quantitative and qualitative data collection and analysis approaches employed in this series of studies. Finally, we discussed the strengths and limitations of the RiPS studies.

Employing Participatory Design (PD) in HRI Research

- PD supports an early involvement of users and an in-depth, mutual communication between researchers and participants.
- Conducting a productive PD session requires multidisciplinary skills with iterative revision to the study design.
- Adopting both quantitative and qualitative approaches with multiple data collection and analysis methods is key to facilitating knowledge discovery from PD.
- Future research is required to scale up PD to in-the-wild, long-term HRI.

Note

1. `https://www.monash.edu/ai-institute/research/flagship-projects/robots-in-public-space`

14

Recruiting Domain Experts and Specific User Populations

Yanjun Yang

HRI systems are often tailored for specific domains, such as surgical applications. Evaluating these systems requires close collaboration with domain experts and targeted user groups. This chapter highlights the challenges associated with recruiting and collaborating with expert groups in HRI research by presenting a project that involves experienced and trainee surgeons [Yang et al. (2024)]. Participating surgeons were invited to assess and compare two control strategies of a robotic laparoscopic holder. Through simulated surgery tasks, the control command choices of participating surgeons were analysed based on their experience levels and the specifics of their procedures.

14.1 Recruitment of Specialised Participants

Chapter 4 provided an overview of how to recruit diverse participant groups, while this chapter focuses specifically on recruiting and collaborating with domain experts, that is, expert sampling.

Expert sampling is a form of purposive sampling [Etikan et al. (2016)] for recruiting specialised participants whose skills are essential for specific HRI tasks, such as, in this case study, working with a surgical robot for suturing. Expert participants are often harder to recruit compared to convenience samples, such as university students, but they provide more insightful responses to targeted research questions.

Engaging domain experts enhances the design and analysis phases of the life cycle of an HRI study (see Figure 1.1 in Chapter 1). During the design phase, experts contribute by refining research questions and hypotheses, suggesting task designs and evaluation criteria, and participating in pilot

DOI: 10.1201/9781032693682-17

tests to further improve the study design. In the analysis phase, their feedback allows researchers to optimise experiments, investigate hypotheses, identify new research questions, and inform follow-up studies.

14.2 Surgeons in a Medical HRI Experiment

In this section, we use a medical HRI experiment [Yang et al. (2024)] as an example to illustrate the common challenges encountered in recruiting and collaborating with expert groups and offer strategies to mitigate these issues. Additionally, we provide guidance on working effectively with domain experts throughout the HRI research process.

14.2.1 Overview of the Medical HRI Project

In minimally invasive surgery (MIS), the surgeon inserts surgical tools and a camera, known as an endoscope, into the patient's body through small incisions. MIS demands close teamwork, typically involving the surgeon and other surgical personnel, such as a camera-holding assistant who positions the endoscope to provide a view of the operating site inside a patient's body for the surgeon.

In contrast, robot-assisted MIS employs a robotic laparoscope holder to replace the camera-handling assistant. This approach avoids issues such as surgeon-assistant communication inefficiency and assistant error, making the procedure more efficient and safer than conventional MIS [Zihni et al. (2018)].

However, the application of new technologies challenges the surgeon both cognitively and physically. Controlling the robotic laparoscope holder can increase the surgeon's workload and distract the surgeon from the primary hands-on operations. Therefore, the medical HRI project aims to develop an intelligent laparoscope control system, composed of a novel foot interface and a shared autonomy framework. This system will allow the surgeon to better focus on the main operation by reducing the laparoscope control workload while still maintaining control authority over the laparoscope [Yang et al. (2020, 2021, 2024)]. The novel foot interface gives the surgeon direct control over the laparoscope, and the shared autonomy framework allows the surgeon to adjust the laparoscope's zoom factor when needed.

In a preliminary analysis, the mechanical design of the novel foot interface was evaluated by 20 participants (average age 21.6 $\pm$ 1.1 years, 2 females, 18 males), who were all undergraduate students from the engineering faculty, through three general control tasks [Yang et al. (2020)]. This study received approval from the university's human research ethics committee as a low-risk study, and informed consent was obtained from all participants.

Beyond mechanical design, another fundamental and critical topic to understand is the best way to coordinate and send commands for a laparoscope control task via a foot interface. To explore this, after the researchers verified the mechanical design of the proposed interface in the preliminary analysis [Yang et al. (2020)], they conducted a study using a laparoscope control task and two surgery-like tasks in a simulator with the target users, that is, the surgeons [Yang et al. (2024)]. Given the task complexity and study goals, the researchers recruited 11 surgeons (6 experienced surgeons and 5 in training), making this study the focal example of this chapter. The study was approved by the university's human research ethics committee, and informed consent was received from each participant.

14.2.2 Challenges in Recruiting and Collaborating with Surgeons in HRI for Medical Applications

This section identifies and addresses the challenges associated with recruiting and collaborating with surgeons across three stages of the HRI project: the pre-recruitment stage, the pre-experiment stage, and the experiment stage. Each stage presents unique difficulties, and understanding these can help researchers better navigate the complexities of working with expert participants.

14.2.2.1 Challenges in the Pre-Recruitment Stage

The pre-recruitment stage is critical because it sets the foundation for the entire project. During this phase, researchers need to establish the project's goals, select recruitment methods, develop effective recruitment materials, and prepare for potential participant dropouts. Getting these elements right ensures a smoother research process in subsequent stages.

Establishing clear project goals involves not only defining research questions but also determining the number of participants needed to achieve sufficient statistical power, as was discussed in Section 10.3. A pre-study can help estimate the required sample size based on a given statistical power value. However, recruiting enough experts often proves challenging. Setting realistic recruitment targets is essential to avoid unnecessary delays and budget overruns. Researchers should also prepare a contingency plan in case the initial recruitment goals are not met.

The team for the study conducted by Yang et al. (2024), which consisted of three academic researchers and two surgeons, initially aimed to recruit 64 participants, based on statistical power considerations and consultations with the surgeon collaborators. However, due to the constraints posed by the COVID-19 pandemic, including significant pressure on the healthcare system, the team was unable to meet this target and instead conducted a smaller study with 11 surgeons.

Common recruitment methods were presented in detail in Section 4.3. In this project, personal contacts and community groups were utilised for recruitment. The collaborating surgeons promoted the study and facilitated the sharing of information among colleagues in their affiliated hospitals.

Recruiting experts from organisations or institutions, such as clinics, through someone who is trusted by that organisation is an effective strategy. This approach underscores the importance of recruiting experts through trusted contacts, as these connections can accelerate access to target participants. Thus, researchers should proactively explore their networks for potential recruitment avenues through collaborators or supervisors in the very initial stage.

Developing effective recruitment materials is the next crucial step once a recruitment method is identified. These materials are often the first, and sometimes only, source of information potential participants have about the study. Recruitment materials should:

1. Clearly explain the study and solve participants confusion
2. Trigger potential participants' interest
3. Outline both benefits and risks

Even when dealing with experts, it's important to use accessible language and avoid jargon, as knowledge gaps can still exist between researchers and participants. Focus on conveying key information without overwhelming participants. For example, in the Yang et al. (2024) study, surgeons who were invited to participate did not have the experience to work with a surgical robot. When explaining the robot motion behaviour, the term "degrees of freedom" was replaced by the phrase, "the number of independent movements or actions the robotic laparoscopic holder can perform". A demo video that visually explained the robot's functions and motions was included to bridge understanding and minimise misconceptions about the study. Explaining a study does not mean overwhelming participants with information. Time is valuable, and only essential and key information should be conveyed to the experts.

Financial incentives can be a good way to motivate participants to enroll in the project. And the standard to follow is that the incentives provided be equal to or proportional to the amount of time and energy required in the study (see Section 4.3). However, for experts, it can often be more effective to appeal to intrinsic motivations.

When recruiting surgeon experts, it is best to assume that most surgeons are altruistic, which means they are more keen to help others and to make a difference to existing knowledge [Clark (2010)]. In recruitment for this case study, the researchers emphasised the potential benefits to the surgical community and outcomes, because altruist individuals may be more likely to volunteer to participate in a study [Newington and Metcalfe (2014)]. In addition, the researchers explained the reason and the importance of hiring experts in the recruitment material. When potential participants feel they can contribute to a study or share valuable insights, they are more likely to join that study [Coyne et al. (2016)]. To motivate potential participants by appealing to their interests in exploring new technologies, the researchers also highlighted the chance to interact with the robotic laparoscopic holder, attracting surgeons eager to engage with cutting-edge research. Once a prospective participant contacts the research team for more information, it is important to always be active and responsive to the communication. This can help the research team to build a professional and positive attitude from the first impression. Letting a prospective participant's message sit in the email inbox for weeks will only consume the participant's patience and can result in negative impacts to the recruitment process.

It is crucial to transparently present all potential risks in recruitment materials, even if the level of likelihood and severity are both low. Some experts may be particularly cautious about such details. Thus, it is recommended that researchers provide such information from the start of their communication with experts regarding the project.

It is unavoidable that some participants will drop out before or during the experiment. Thus, researchers need to prepare for this in their study design and experiment plan. Participants might drop out due to a time conflict or not feeling safe or comfortable to participate or continue. Understanding the reasons behind dropouts, when possible, can inform future improvements in recruitment and study design.

14.2.2.2 Challenges in the Pre-Experiment Stage

The pre-experiment stage is a vital period before conducting the large-scale experiment for refining research questions, improving the experiment design, determining evaluation criteria, and conducting pilot studies. Expert input during this stage can improve the quality and relevance of the study.

Chapter 2 introduced two main types of HRI studies: exploratory and confirmatory. In the preliminary study, Yang et al. (2020) focused on validating specific hypotheses about control strategies for a robotic laparoscopic holder through a confirmatory approach.

A good control strategy is a key part of developing a suitable foot interface for the laparoscope-holding robot in addition to proper mechanical design. An interface needs to control at least four degrees of freedom (DoF) of a laparoscope to provide suitable focus and orientation of the target site view. The two most common control approaches are decoupled and hybrid control. Decoupled control actuates a single degree of freedom at a time, known as decoupled commands. Hybrid control allows separate actuation of each degree of freedom, as well as simultaneous actuation of multiple DoF, known as coupled commands. When proposing each of these designs, designers commonly make the following two hypotheses:

- Decoupled control makes predicting the final view orientation of the laparoscope tip easier for the operator [Yang et al. (2020), Abdi et al. (2016), Voros et al. (2010)].
- Hybrid control is more efficient compared to decoupled control [Huang et al. (2019), Sanchez et al. (2021), Zuo et al. (2022)].

The efficiency and predictability hypotheses underlying the choice of control configuration in [Yang et al. (2020)] have not been tested in previous research, which shows a research gap in this domain. Thus, [Yang et al. (2020)] aims to validate these two hypotheses.

In the case study, discussions with supervisors and consultant surgeons revealed an additional, unexamined area: How do surgeons choose between coupled and decoupled commands in surgery-like tasks, considering task requirements, task phases, and personal experience? This insight led Yang et al. (2024) to expand the study's scope to include both exploratory and confirmatory research objectives.

Experiment design involves setting up the environment, defining tasks, determining session frequency and duration, and selecting evaluation criteria, as was detailed in Chapter 3. All these factors need to be thoughtfully considered for expert participants to increase the reliability of the experiment results. Expert collaboration also ensures these elements are tailored to realistic scenarios.

One of the essential purposes of all HRI studies is to understand how people interact with robots in simulated or real-life scenarios. Collaborating with experts helps researchers to set up a more realistic simulated experiment environment. For example, in the Yang et al. (2024) project, the RA-MIS simulation layout (Figure 14.1) replicated the setup used in the operating room. The endoscope was inserted from the surgeon's side, and both the endoscopic view and the foot interface overlook view were displayed on a monitor placed in front of the participants. The surgery simulator was a $300 \times 220 \times 100$ mm^3 open box covered by a board; that is, participants did not have a direct view of the operating site and had to rely on the laparoscope view. An LED light was attached beneath the cover to provide sufficient and stable lighting. As suggested by the collaborating surgeons, two L-bracket holders with 6-mm openings (Figure 14.2) near the simulator were used to simulate trocar sleeves.

The experiment included two task types: a pure laparoscope control task and a set of surgery-like tasks. The pure laparoscope control task, commonly used to test foot interfaces [Yang et al. (2020), Huang et al. (2021)], was designed to validate the efficiency and predictability hypotheses of the control strategies. The surgery-like tasks, chosen to examine the surgeons' usage patterns, were designed with specific considerations: ensuring the tasks required laparoscope-tool coordination skills and maintaining a realistic context. Therefore, two candidate tasks, the Peg Transfer task and Suture with Intracorporeal Knot task, were selected from the Fundamentals of Laparoscopic surgery program [Vassiliou et al. (2010)], which is a widely recognised training program by surgeons that contains tasks for evaluating the basic skills required to perform laparoscopic surgery. Another candidate task, the needle threading task, which is commonly used in commercial laparoscopy simulators [LAPARO Sp. z o.o. (2021), Kroton Medical Technology (2023)] was also considered.

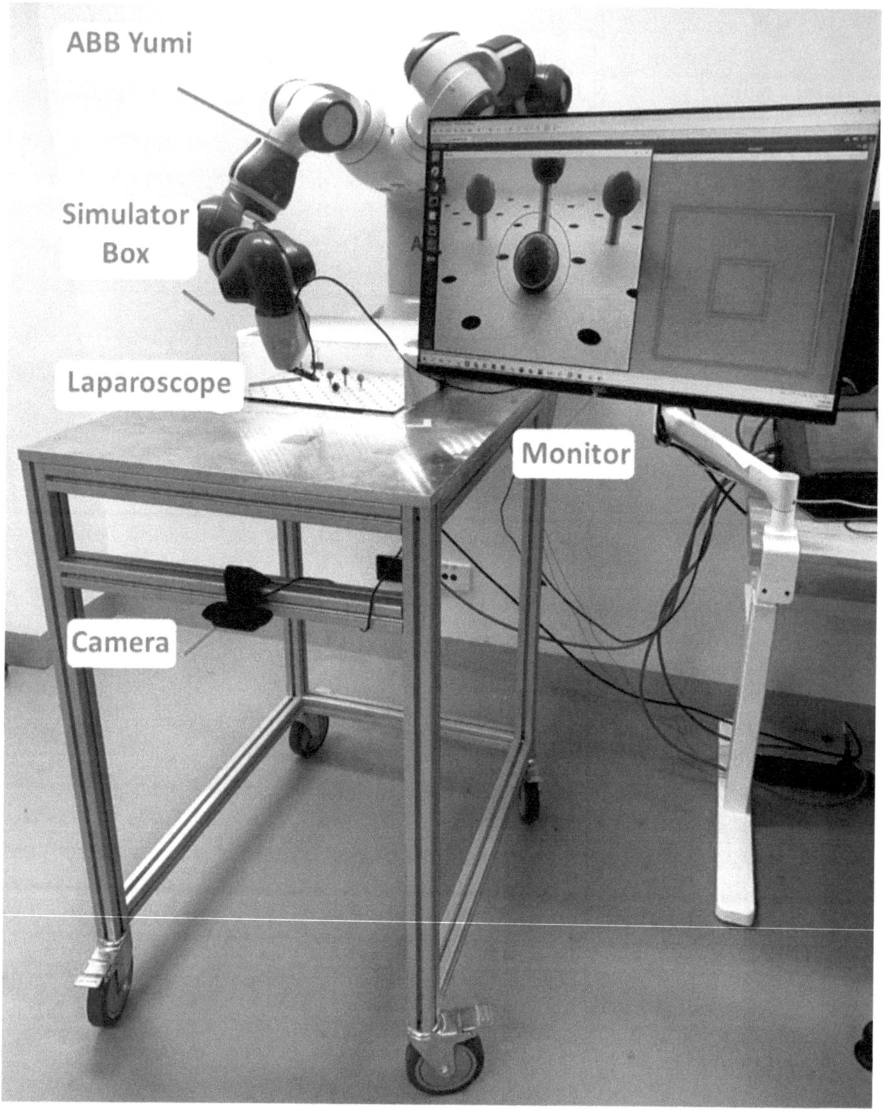

FIGURE 14.1
The laparoscope-holding robot system used in the experiment consisted of a YuMI robot, a surgery simulator, and a monitor. [From Yang et al. (2024)].

The suturing task was excluded due to safety concerns associated with replacing the suturing set (containing needle), as well as the longer time required to complete the task, which could have extended the experiment duration significantly. The collaborating surgeons advised that keeping the experiment duration within 1.5 hours was crucial because the experts

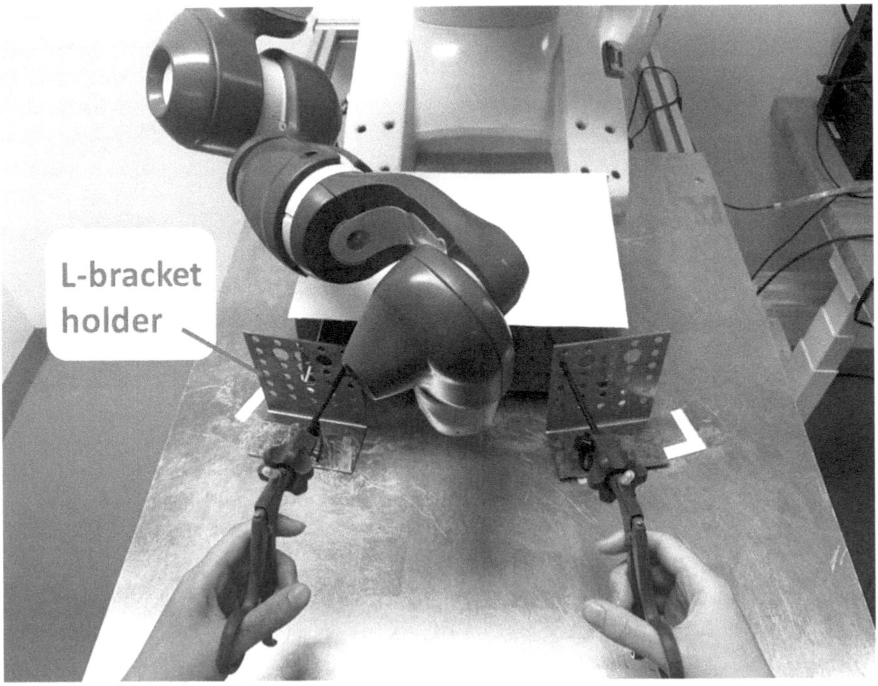

FIGURE 14.2
Two L-bracket holders with 6-mm openings provided a pivot point for the inserted surgical tools, as suggested by collaborating surgeons. [From Yang et al. (2024).]

usually had very tight schedules and most of them used their break times during the weekdays for taking part in the study. Thus, a study duration longer than 1.5 hours could make it difficult to surgeons. This demonstrates the importance of selecting a suitable session frequency and duration. Researchers not only need to avoid the impact of the novelty effect, which was introduced in Section 3.2.2, but also need to consider the convenience and preference of the participants.

The two selected surgery-like tasks were modified to better fit the study's research questions. The Peg Transfer task was adapted into a Pick&Place task. The experiment setup consisted of five poles, including one central pole and four others arranged around it, labelled P1, P2, P3, and P4 in a clockwise direction [Figure 14.3(a)]. The central pole has three triangular objects arranged in a stack. A surgical grasper and curved forceps were provided as the available tools. The participant was instructed to remove these triangular objects from the central pole using one tool, transfer the object to the other tool, place the object on one of the surrounding poles,

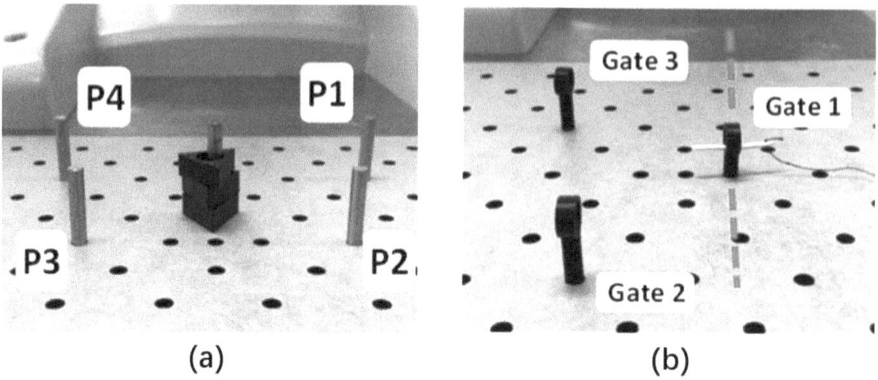

FIGURE 14.3
Two surgery-like tasks: (a) Pick&Place task setup, and (b) Lead-through task setup. [From Yang et al. (2024).]

and then repeat the task until all three objects were placed on the surrounding poles.

The needle-threading task was modified to a task named the Lead-through task. A needle was placed in Gate 1 at the beginning [Figure 14.3(b)]. The participant was asked to use the provided surgical tools to hold the needle all the time and thread it through three circular gates (Gate 2–Gate 3–Gate 1). The participant could choose the toolset of two needle holders or one needle holder and one curved forceps based on individual preference.

Measurement criteria can influence the results. In HRI studies, participants are typically instructed to perform the task as fast as possible while trying to meet all criteria and minimise errors. Criteria selection is a crucial aspect of experiment design, particularly where experts can contribute. The level of difficulty of the two surgery-like tasks in the case study was not very high for the recruited surgeons, as these tasks involved basic skills. Without appropriate constraints, it would be challenging to observe differences in usage patterns among the surgeons, and the results might not accurately reflect their behaviour in real-world applications. To address this, four safety criteria were introduced, based on advice from the consultant surgeons, to ensure that participants behaved realistically within the simulation environment:

- *The number of times the object droped*: Surgeons may inadvertently drop the object and leave it inside the patient body. This is dangerous in real surgery.

- *The total time that the held object was outside of the laparoscope view*: Keeping the held object outside the view is dangerous as users cannot see the object, which decreases their situational awareness.

- *The number of times the tool hit the board*: In the setup, the baseboard was analogous to the patient's tissue and the pole/gate was the operation site. Hitting the patient with the tools should be avoided.

- *The less error-prone approach*: Participants were asked in questionnaires to identify the approach they found less error-prone if implemented in the actual surgery.

As an example, in the Pick&Place task, the pilot study in [Yang et al. (2024)] revealed that the triangle could be placed onto a pole without adjusting the laparoscope. However, with the addition of the safety criteria, surgeons were required to keep the triangle object within the laparoscope's view at all times. This necessitated frequent adjustments of the laparoscope during the task, aligning the workflow more closely with real-world surgical scenarios.

When designing experiments for experts or specific users, it is recommended that a combination of quantitative and qualitative evaluation methods be used. Given the challenges of recruiting experts, researchers should aim to collect diverse data types from each participant. In addition to common task performance metrics (see Section 5.2.1) such as task completion time or error rate, it is valuable to study human-related constructs (see Section 5.1.2) to understand changes in participants' physiological signals during collaboration with a robot. For instance, researchers can equip participants with biomedical sensors to monitor dynamic heart rate. The entire experiment can also be recorded, with participant consent, for observational purposes. Observation (see Section 11.1.1) serves as a qualitative measurement method that minimises researcher intervention and captures the participant's natural responses in the HRI study.

Pilot studies play a critical role in identifying potential deficiencies with the research design prior to the main study. More details about pilot studies can be found in Section 6.1. In the expert-focused pilot study, the researchers prioritised the clarity of the testing questionnaire and providing an accurate estimate of the total time commitment for participants, ensuring that the study design was both practical and informative.

Estimating the total time required for an experiment can be complex and involves more than just adding up the time spent on each task. Timing begins as soon as the participant enters the lab and includes several additional components: completing consent forms and other paperwork, reading or receiving an introduction to the experiment, participating in training sessions, performing the tasks, resetting the experiment environment for repeated trials, filling in subjective surveys if required, handling any potential failures, and allowing for breaks during longer sessions. It is essential to strike a balance between the ideal session duration and frequency and the maximum time participants can reasonably commit, especially for those with tight schedules.

14.2.2.3 Challenges During the Experiment Stage

The experiment stage is where researchers put their plans into action. Working with experts requires additional considerations to build rapport and trust. Experts and users from specific groups are difficult to recruit. When a participant is willing to enroll in a study, it is recommended that researchers take this chance to give the participant a positive experience with the experiment design and schedule arrangement, such as improving the convenience and flexibility of the experiment. By doing this, researchers may have a higher chance to add these participants to an experts pool and ask them for their availability and willingness to take part in future studies that would require their expertise again.

Flexibility and convenience in accommodating study schedules are essential to make participants feel respected and can aid in recruitment efforts [Lindsay et al. (2021)]. Expert participants have their own schedules, and participating in the study may not be their priority. Offering various time slot options, such as mornings, lunch breaks, after work, or weekends, can provide the flexibility needed to fit the experiment around a participant's schedule. One recommended approach is to provide the participants with a publicly accessible calendar that is updated in real time, allowing them to view available time slots and make adjustments to a booked session if emergencies arise. In this case study, the researchers shared a Google appointment calendar with all participants for this purpose.

It is also important to leave sufficient time between slots to prepare and reset the experiment setup. Respect participants' time by ensuring they do not have to wait, and have everything ready before their arrival so the experiment can start promptly. Additionally, providing participants with two to three reminders about their upcoming session, including one on the day before, helps ensure compliance and reduces the risk of missed appointments.

Convenience also involves making it easy for participants to reach the trial location. This case study was conducted in a university laboratory to allow convenient access to robotic equipment. Providing participants with information about the best parking locations or recommended public transport routes the night before the experiment can significantly reduce stress. In this study, the researchers also provided a detailed map with photo instructions to guide participants from the car park or bus station to the laboratory. Offering parking vouchers, when possible, is an effective way to further enhance goodwill. Finally, it is important to ensure that participants know how to contact the experimenter and be responsive and available through those channels to address any queries in a timely manner.

Providing food, snacks, and water during the experiment is a simple yet effective way to build rapport, especially when the experiment is lengthy or scheduled during common meal times. In the case study, the researchers provided each participant with a protein bar and a bottle of water after the experiment. During the COVID period, they also supplied hand sanitiser to maintain hygiene standards.

Professionalism is crucial for building trust with participants: Researchers should be thoroughly familiar with the study protocols and maintain a positive and respectful demeanour throughout the experiment. This helps establish credibility and trustworthiness, encouraging experts to engage fully with the study. It also helps in minimising human error when setting up the experiment and during data collection. In addition, transparent communication is important and researchers should be prepared to answer questions from the participants. In the case study, some expert participants showed great interests in the research project and future development plan. Some participants asked for clarification or a repeated statement of the task process instructions. In addition, they asked questions regarding the rationales of some task design choices. Being prepared for these questions demonstrates the professionalism of the research team to a participant and establishes trust with the participants.

Conducting the experiment daily can be tiring for researchers in the long run. However, maintaining a positive attitude is crucial, as it helps participants feel relaxed and comfortable, fostering a sense of trust and credibility between researchers and participants. This can be achieved by sustaining enthusiasm, offering positive feedback, and providing encouragement during or after the tasks. It is also recommended that researchers plan the experiment sessions with breaks to allow themselves to rest and be better prepared.

Finally, in the post-study phase, maintaining communication with experts or specific users after the study, such as sharing updates on research progress or summarising study findings, can help sustain interest and foster long-term relationships. Acknowledging participants' contributions in publications and keeping them informed on study outcomes are important steps in maintaining goodwill and encouraging future participation.

14.2.3 Discussion

We will now discuss the strengths and limitations of the medical HRI case study in the context of recruiting and collaborating with domain experts.

14.2.3.1 Strengths

Recruiting surgeons to participate in the robotic laparoscope holder project allowed the research team to validate the efficiency and predictability assumptions of the decoupled and hybrid control approaches directly with end-users. Additionally, observing the participating surgeons' usage patterns during surgery-like tasks provided valuable insights into how surgeons interact with the laparoscope-holding robot based on task phases, their level of experience, and the speed of laparoscope movement. Studying these usage patterns is crucial as it reveals implicit needs that might not be captured in the initial design requirements.

The involvement of professional medical experts lends greater credibility to the study design and resulting conclusions. Without participants with a relevant medical background, the findings can have limited validity. By utilising surgery-like tasks and incorporating feedback from surgeons, the study was able to develop essential guidelines for the foot interface design tailored to actual users' needs. While the results specifically inform the design of foot interfaces in laparoscopic control, the experimental protocol can be adapted for other HRI studies, particularly those involving tool-manipulating robotic arms in surgical and industrial applications. Such studies often require participants to demonstrate strong user-tool coordination skills in addition to maintaining control.

14.2.3.2 Limitations

Despite the aforementioned strengths, this project had several limitations. The primary limitation was the small sample size, which resulted in low statistical power, with power values below the recommended 0.8 threshold. To achieve statistically robust results, the project would have had to recruit at least 64 surgeons. Thus, if recruiting domain experts or specific users is essential for a study, it is crucial to begin the pre-recruitment stage and participant recruitment as early as possible. Further, a statistical analysis alternative to the traditional NHST, as was discussed in Sections 10.2 and 10.3, may be adopted when recruiting more participants is not feasible.

Another limitation in the medical HRI project was the lack of objective measurements for assessing surgeons' fatigue and cognitive load. These aspects were only evaluated through participant feedback and questionnaires. Future studies could incorporate more quantitative methods, such as electromyography (EMG) for fatigue measurement and electroencephalography (EEG) for assessing cognitive load changes. Combining subjective and objective measurements could provide a more comprehensive analysis of participant performance.

In addition, the participants' mastery of the foot interface was not formally tested due to time constraints. In our experiment, each participant received a 5-minute training session to explore all foot commands and speed levels of the robotic laparoscopic holder. Participants could end the training session once they felt comfortable with the interface before moving on to the formal tasks. While this training approach is commonly used in similar studies [Huang et al. (2019, 2020)], establishing a bench-mark for participants' mastery of the HRI interface could help reduce bias and improve result analysis. However, including performance evaluations during training can significantly increase the experiment time and make recruiting domain experts even more difficult. Researchers are encouraged to develop efficient assessment methods that quickly gauge mastery or consider separating the training and task sessions into different time slots or days.

Furthermore, this experiment assessed the participating surgeons' performance when they used the new interface for the first time. It is not clear if these results would remain consistent with extended use and in a more time-consuming surgery-like task. Therefore, future work should conduct larger and longer-duration experiments to address these limitations. Moreover, the surgeons were invited to participate in the experiments in a lab setup, which may not fully replicate the spatial and task context compared to in-situ testing in a hospital operating room. Future experiments that incorporate a more realistic operating context would allow evaluation of the new interface with increased ecological validity. However, such evaluation would require careful considerations to develop the experiment in a way that minimises disruptions to the normal workflow of the site, and ethics approval from the collaborating medical facilities would need to be acquired in addition to an ethics approval from the university board.

Lastly, the experiment environment could be enhanced for a more realistic experience and the validity of the collected data. The goal of recruiting domain experts or specific users is to gather genuine feedback and observe authentic interactions with the robot. Therefore, it is the research team's responsibility to create a realistic simulation environment that closely resembles the participants' actual working conditions. In this case study, an in-house simulator was used; a commercial surgical simulator may offer more ergonomic tool insertion port positions and a trocar-like feel with rubber wrapping around the insertion site, providing an experience closer to real MIS.

14.3 Summary

In this chapter, we reviewed a medical HRI project as an example to illustrate the challenges in recruiting and collaborating with domain experts, specifically surgeons. The chapter began by emphasising the importance of involving domain experts and outlining the main challenges encountered at three key stages of the study: pre-recruitment stage, pre-experiment stage, and experiment stage. We then discussed the specific challenges and collaboration strategies for each stage. In the pre-recruitment stage, challenges included setting statistically valid and feasible recruitment goals, choosing effective recruitment methods, and designing materials that resonate with the target participants. In the pre-experiment stage, the main challenges involved refining research questions and designing experiments, including task creation, evaluation criteria selection, and conducting pilot studies. During the experiment stage, the primary challenge is building trust and rapport between the research team and participants. This chapter aimed to highlight the complexities of recruiting and collaborating with domain experts, with specific details of the medical HRI project serving as a context for the discussion and recommendations. This case study reflected on discussion in Chapter 4 on participant recruitment.

Recruiting Domain Experts and Specific Users in HRI Research

- Before the recruitment stage, the research team should set recruitment goals early that balance statistical validity with the feasibility of recruiting the required number of participants. Unlike convenience samples, recruiting domain experts often requires different strategies, such as leveraging professional networks through trusted contacts and collaborating with experts to connect with potential participants and streamline the recruitment process.

- Involving domain experts in the study design helps refine research questions, develop tasks, and create scenarios that accurately reflect real-world applications, enhancing expert engagement. Experts can also help establish suitable evaluation metrics that capture natural and genuine performance aligned with actual work behaviours. Accurately estimating total experiment time is crucial, because experts typically have demanding schedules, and respecting their time is essential.

- Building trust and rapport with domain experts and specific users during and after the experiment is critical. It can encourage experts to participate fully and increase their willingness to take part in future studies that require their expertise again. Some useful strategies include offering flexible scheduling options; providing logistical support such as transport instructions, parking vouchers, and snacks; and maintaining a positive and professional atmosphere. Providing positive feedback and encouragement fosters a welcoming and engaging environment.

15

Simulated Humans and Sim2Real Studies

Simulation environments allow for efficient data collection and rapid development of HRI systems. They are especially useful when there are safety and/or resource constraints for conducting in-person user studies. However, the sim-to-real (sim2real) gap remains an unsolved issue in HRI research, considering the difficulty of incorporating contextual, interactive, and human factors in the simulation environment. This chapter examines an example study where a drone landing assistive system for novice pilots was designed and evaluated, first in simulation and then in the physical world.

15.1 The Role of Simulations in Human-Robot Interaction

As was discussed in Chapter 4, recruitment of human participants can be challenging due to limitations in equipment availability, access to end-users, and other factors. With the advent of deep learning methodologies in robotics and HRI, a large corpora of HRI data may be needed, amplifying recruitment challenges.

Recently, many deep learning algorithms in robotics have employed a two-stage pipeline: The algorithm is first trained in simulation, where many experiments can be conducted rapidly and without safety constraints, and then subsequently deployed on the physical robot. The success of this approach is strongly impacted by any differences between the simulated and physical environment, known as the *sim2real gap* [Höfer et al. (2021)]. While early solutions to the sim2real gap focused on developing higher fidelity simulators, recently, *domain randomisation* has been proposed as a powerful solution to the sim2real gap [Muratore et al. (2022)]. Domain randomisation, rather than attempting to faithfully replicate exact physical conditions, samples from a variety of physical conditions and trains the robot to perform well across a population of conditions (e.g., variation in environmental

DOI: 10.1201/9781032693682-18

properties and robot properties). Then, when the robot is deployed in the physical world, the physical world will be a sample within the distribution the robot has already been trained on, resulting in a successful zero-shot transfer from simulations to the physical world [Muratore et al. (2022)].

When training interactive robot policies in simulation, the performance of the learned policy is strongly influenced by the diversity of interaction experiences the robot will encounter during learning [He et al. (2023), Ah Sen et al. (2022)]. These experiences, in turn, depend upon the accuracy and quality of models used to control the behaviour of the simulated humans inhabiting the simulation environment and with which the robot will interact during learning [Markkula and Dogar (2024)]. In this context, choosing or crafting an appropriate model of human behaviour is key. In this chapter, we detail a case study investigating how these ideas can be applied to HRI.

15.2 Case Study: Assisting Human Pilots with Drone Landings

As an example of adopting sim2real approaches in HRI research, we review an illustrative study that used simulated humans to train a reinforcement learning policy to assist human pilots with drone landings [Backman et al. (2023)].

15.2.1 Study Aims and Methodological Motivations

Unmanned aerial vehicles (UAVs) are increasingly being deployed in many different applications, including search and rescue [Perez-Grau et al. (2017)] and infrastructure inspection [Sa et al. (2015)]. However, when UAV tasks are mission-critical or in highly dynamic environments, human teleoperation is generally still preferred to fully autonomous flight [Shaqura et al. (2018), Perez-Grau et al. (2017)]. While UAVs are valued for their high maneuverability in three-dimensional (3D) space, this capability creates challenges for teleoperation, where the pilot's perceptual and control abilities will be put to the test [Smolyanskiy and Gonzalez-Franco (2017), Kim et al. (2020)]. As a result, shared autonomy systems that could assist human pilots with challenging teleoperation tasks, such as landing, would be highly desirable to enable a greater diversity of operators to successfully pilot drones.

An automated drone assistant must successfully answer two questions: (1) What is the intent of the user? For example, does the pilot want to initiate landing? and Where does the pilot wish to land? (2) What kind of assistance would be useful? For example, does the pilot need assistance with positioning due to impaired depth perception? The answer to these questions will depend on the current context and capabilities of each pilot. In this project, the aim was to develop an intelligent assistant that would be able to recognise the pilot's intent and needed assistance online during shared autonomy operation, creating a single policy that could robustly provide appropriate assistance across a range of pilot preferences and capabilities.

To implement the drone assistant, a two-stage design process was followed. First, the assistant was trained using reinforcement learning in a simulated environment, using a parametrised *user model*. The parameters of the user model aimed to capture the diversity of user intent and capability, to enable training of a single policy suitable for diverse users, purely in simulation. Training in simulation allowed for a large dataset to be collected, without the limitations imposed by participant availability and human and robot safety. Second, the trained assistant policy was deployed on a physical drone and evaluated in a physical HRI study with diverse participants.

15.2.2 User Model and Simulation Training

To train the assistant policy in simulation, a model of the human pilot was used which captured how the human pilot would behave in response to environmental context and automatic assistances. The idea of domain randomisation, previously proposed for physical properties [Muratore et al. (2022)], was extended to HRI by creating a parametrised model. The parameters of the model captured the key dimensions of variability in human behaviour. It was desirable for the model to be parsimonious (i.e., have few parameters) to minimise the amount of samples and speed up training.

In the case study, a human pilot was simulated as a state machine with two states: approach and descent. In the approach state, the simulated pilot flew towards the intended target, while in the descent state, the simulated pilot would attempt to land the drone on the intended target. Four types of human variability were modelled: perceptual capability, acceptance of assistance, preferred speed, and teleoperating capability. In the approach state, simulated users planned their trajectory based on where they perceived the landing target to be, as determined by their perceptual capability. For simulated pilots with perfect perceptual capability, the perceived target was co-located with the true target, while degraded perceptual capability corresponded to increasing offsets between the true and perceived target. As the user was working in shared autonomy with the drone, the assistant would correct the user input. The user might be more or less receptive to this input from the assistant. This was modelled with the acceptance of the assistance parameter. Once users began executing their planned trajectory, they might have different preferences about how fast they wished to proceed. This was modelled with the preferred speed parameter. Finally, each user might be more or less skilled with the teleoperation interface. The researchers modelled the user control as a proportional controller, with differing controller gains corresponding to

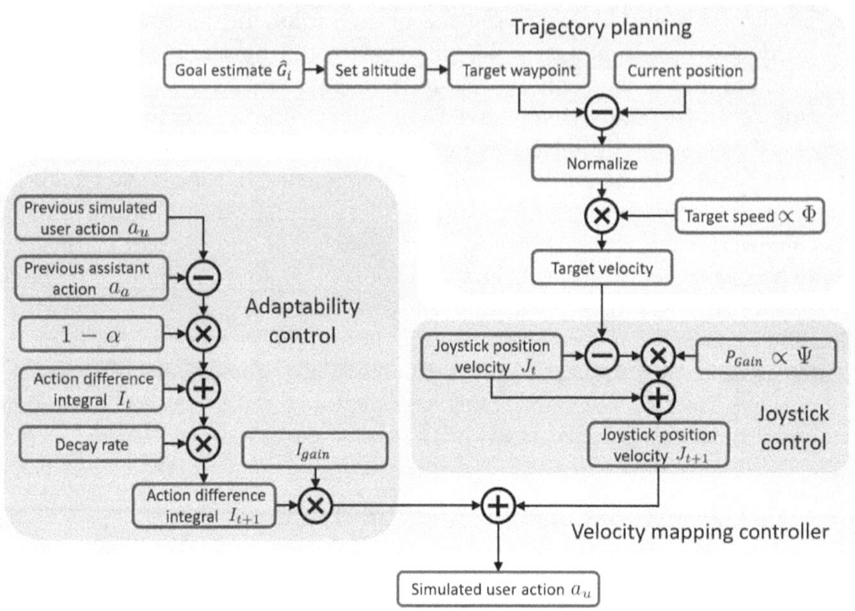

FIGURE 15.1
Simulated user model. [From Backman et al. (2023).]

differences in teleoperating capability. The simulated user model is summarised in Figure 15.1.

Given this user model, the assistant policy was trained in simulation, using domain randomisation over simulated users and the environment. The policy network used the TD3 architecture [Fujimoto et al. (2018)], but with added Long Short-Term Memory (LSTM) layers in both the actor and the critic to capture the time history of states and actions. The reward function penalised the landing error, any velocity at landing, and any differences between the human and the assistant actions.

During training, the critic was aware of the true goal position of the simulated user. This information was not available to the actor during the interaction; the actor had to infer the landing target of the user through the user's actions and the environment context. But an important advantage of using simulated users during training was that privileged information (i.e., the known target location) could be provided to the critic, which could substantially improve performance and training speed.

15.2.3 Physical Experiments

After training a policy in simulation, a physical experiment was conducted to evaluate how well the policy transfer red to physical human-robot

FIGURE 15.2
Experiment setup [Backman et al. (2023)]. The left panel shows the experiment facility with the motion capture system and nine landing platforms. The right top panels show the drone used for the experiments. The right bottom panel shows the positioning of the pilot with respect to the drone flight arena and target platforms.

collaboration. In a large laboratory space, the researchers set up an array of landing platforms as potential landing targets. Participants were instructed to land at a particular platform and used a standard joystick drone teleoperation interface to land the drone. The experiment setup is illustrated in Figure 15.2. The study protocols were reviewed and approved by the university's human research ethics committee.

The automated assistant used the information from its onboard cameras to detect potential landing platforms and information from the motion capture system to estimate its current position and velocity. The command from the user input, estimate of current position and velocity, and output from onboard perception were concatenated and provided as input to the trained assistant policy. The assistant did not receive any information about what the target platform was. This had to be inferred from the human pilot's control inputs. Finally, the outputs of the automated assistant's policy module and the user commands were averaged together to provide a target velocity input to the drone flight controller. No additional training or fine-tuning was performed on the automated assistance; that is, adopting zero-shot transfer from the simulated user training. The overall system architecture is illustrated in Figure 15.3.

Each participant performed the task in two conditions: assisted or unassisted. The order of conditions was random. In each condition, participants were tasked with landing on ten randomly selected platforms. As can be observed in Figure 15.2, the overhead lights were turned off so that participants could not use the shadow of the drone to help them perceive depth, creating challenging perceptual conditions.

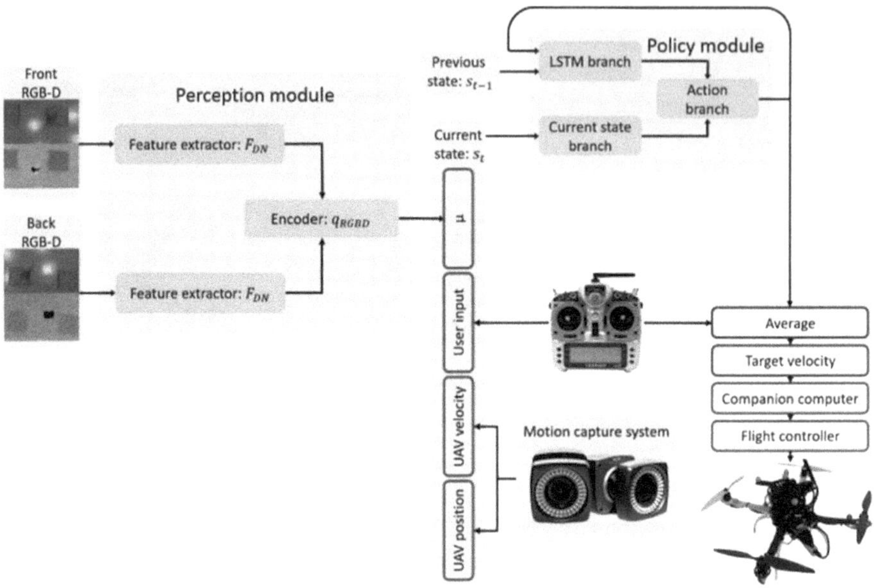

FIGURE 15.3
System architecture. [From Backman et al. (2023).]

Twenty-eight participants were recruited to the experiment. The majority had little or no drone piloting expertise, while a small number had more extensive previous expertise and training. Participants received no piloting training prior to the start of the experiment.

15.2.4 Key Findings

The task performance was evaluated on two task metrics: landing error, the distance between the actual landing location and the centre of the target platform, and (2) the success rate (Did the drone successfully land on the target platform without crashing or toppling over?). Figure 15.4 summarises the results.

As can be seen from Figure 15.4, the landing error was substantially reduced in the assisted condition, while the success rate was close to 100% across all participants. While improvements in landing error and success rate were largest with the least proficient pilots, all pilots benefited from the assistance. The combination of pilot and assistant enabled all pilots to perform at a similar level, removing any relationship between task success and pilot proficiency.

These results highlight the possibilities of the use of simulation for enhancing human-robot collaboration while reducing the requirements for time- and resource-intensive human data.

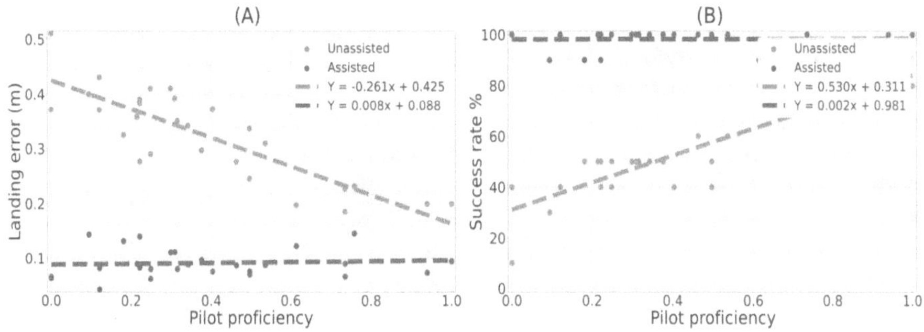

FIGURE 15.4
Results summary. [From Backman et al. (2023).]

15.3 Discussion

In this section, we discuss the strengths and weaknesses of using simulation environments in HRI research, as illustrated in the case study.

15.3.1 Strengths

Deploying human models in lieu of human data collection offers several advantages. Large amounts of training data can be collected quickly and more easily, which is particularly valuable for deep learning approaches. Concerns about participants' physical and psychological safety, particularly while the robot is learning, can be ameliorated. With appropriately selected model parameterisation, participant diversity can be simulated without requiring extensive recruitment efforts. As was discussed in Section 3.2.1, online and simulator environments can be valuable mediums for HRI experiments to facilitate the collection of large and diverse data. While this case study focused on using simulated environment and human models in HRI, in Chapter 16 we will exemplify HRI experiments with online, crowd-sourced participants.

15.3.2 Limitations

On the other hand, a key limitation of simulation-based approaches is the validity of the human model. If the model does not accurately capture the key characteristics and variations of human behaviour during interactions, the learned policy is unlikely to transfer from simulation to reality. In the case study discussed in this chapter, domain-specific knowledge about likely variations in human perceptual and control capabilities during drone piloting was deployed to create a task-specific human model. However, it is not clear how such models can be developed for other HRI tasks. As was discussed in Section 3.2.1, developing realistic models of human behaviours to enhance simulators for HRI research remains an open challenge.

While the task-specific model enabled successful policy transfer from simulation to physical experiments, it still does not capture all dimensions of user variation, even within the specific task. For example, even within the short-duration experiments, some participants were observed to change their behaviour over the course of the experiment with the assistant. Some participants engaged in testing behaviour, purposefully pushing the drone to see what the assistant would do. Other participants increased their reliance on the assistant as they better understood its capabilities. These types of time-varying human behaviours, individual differences, and adaptation were not captured by the user model.

15.4 Summary

In this chapter, we reviewed a case study deploying simulations and user models to facilitate the development and evaluation of human-robot collaboration policies. As illustrated in this case studies, simulation can be a powerful tool to address the need for data in computational models for HRI. However, to address the Sim2Real gap, it is essential to accurately model human behaviours.

Human Models and Sim2Real in HRI Research

- Human models enable rapid data generation, supporting learning of HRI robot policies in simulation and decreasing the challenges associated with human data collection and participant recruitment.
- Simulation-based approaches require an appropriately parameterised human model, One that captures the main characteristics and dimensions of variability of human interactive behaviour.
- Human models are not perfect and may not capture less common or more long-term variations in human behaviours.

16

Leveraging Online Studies to Expand Diversity, Sample Size, and Response Quality

Aimee Kathleen Allen

In-person user studies are essential for understanding human behaviours and perceptions towards robots, under a given context. However, accessibility to hardware, robots, and experiment sites can severely limit the diversity and number of participants for an in-person HRI study. Online studies using video stimuli can help to both expand the participant population, because of the reduced recruitment and attendance costs of online studies, while also increasing the diversity of participants by including geographies and subgroups beyond those who are able to access a single physical location. Such benefits have led to an increased adoption of online HRI studies in recent years.

On the other hand, online studies may suffer from a reduced quality of responses due to the lack of direct interaction and co-presence of the participant and robot in a spatial context. This chapter illustrates how to effectively leverage online studies to mitigate some of the perceived weaknesses and generate high-quality responses. An online study examining human perceptions of robots relating to the sounds produced by robots is presented as a case study. This study collected data from a large population of participants from multiple countries, varied demographics, and a range of experience levels with robots. The responses obtained had high completion rates for both quantitative and qualitative questions, enabling a diverse set of quality responses to be collected. The pros and cons of online studies, as exemplified by the case study, are discussed, including recommendations to assist researchers in designing future online HRI studies.

DOI: 10.1201/9781032693682-19

16.1 Why Do Online Studies?

Online studies have different advantages and weaknesses to in-person HRI experiments, as was discussed in Section 3.2.1. Motivations for running a study online include the following:

- *Increase participant diversity*: In-person studies are restricted by a participant's geographic location, as was discussed in Section 4.3.2. Online studies may make it easier to recruit diverse cohorts including those who are geographically distant from the research team, physically impaired, or introverted.
- *Increase participant numbers*: The lower cost of participation (both to researchers and participants) makes it easier for to recruit participants.
- *Provide identical stimuli for participants*: The same set of video stimuli are shown to online participants, which reduces potential influence of confounding factors beyond the study variables.
- *Provide anonymity for participants*: When participants can say what they think without consequence, there will be a reduced social desirability bias, as was discussed in Section 10.1.
- *Reduce resource requirements*: Online studies have lower resource requirements than in-person studies, including both financial costs and researcher time.
- *Allow for exploratory studies*: Online studies can be used to evaluate the number of participants required for a follow-up in-person experiment.

The following section examines a case study experiment that demonstrates how to leverage online studies for HRI research.

16.2 Case Study: Human Perception of Consequential Sounds Produced by Robots

As an example of the techniques used to leverage online study designs, a case study research project is presented in this chapter. This HRI research study, entitled "Effects of Consequential Sounds on Human Perception of Robots" [Allen et al. (2024b)], investigated how the sounds produced by robots as they move and operate are perceived by people and the resulting implications for HRI and robot adoption in shared spaces. Further details are available on the project website [Allen et al. (2024a)].

As robots perform their normal movements and operation, their actuators generate unintentional, audible machine "noises" known as "consequential sounds". This case study looked at what effect the presence of these "consequential sounds" had on the human perception of robots, providing evidence that consequential sounds create more negative human perceptions of robots, thus affecting participants' desire to co-locate with robots.

The experiment was run online using the Qualtrics platform and took participants roughly 30 minutes to complete. A parallel mixed-methods approach was used, asking participants to respond to both quantitative and related qualitative questions (see Section 11.2 for an explanation of the benefits of this approach). The experiment consisted of a pre-questionnaire of demographic and robot-familiarity questions, sound tests to verify participants' ability to hear sounds, five trials each containing a 20-second video of a robot and per-trial questions, and a final post-questionnaire. For each trial, participants watched videos of one of five robots as they moved and produced consequential sounds, before providing their feedback on each robot, including both Likert scales and qualitative responses. A between-participants study design allowed comparison between participants who experienced the "consequential sound" (CS) stimuli, and those in a silent "no sound" (NS) control condition. The robots used in this study are shown in Figure 16.1.

16.2.1 Motivation and Benefits from Running This Experiment Online

This experiment was intentionally designed to be run as an online experiment. The research topic of this case study, "sound in HRI", is a largely underresearched area, so there is less existing research to support its premises. In cases such as these, the ability to collect exploratory qualitative data can be important, and the number and diversity of participants are especially important.

The benefits of conducting this case study as an online experiment included:

- *A large sample size (volume of responses)*: This was made achievable by reducing costs for researchers (since they were not required to be present during the experiment) and participants (since they could participate from anywhere with Internet access at a time that suits them).
- *Reflective of a diverse population*: Having a diverse population leads to a good range of responses.
- *More accurate controls*: For example, all participants saw identical stimuli in the videos.

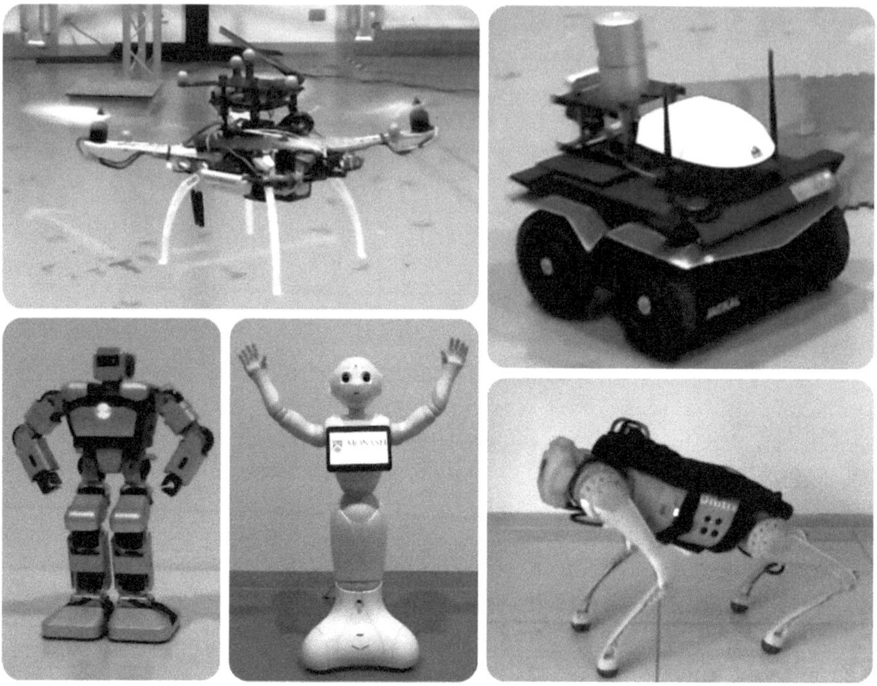

FIGURE 16.1
Still frames of each robot from videos in the case study online experiment. Participants watched a video of each robot and answered questions regarding their perception of that robot.

- *Increased anonymity for participants*: No names or identifiers were collected, so even unique ways of writing qualitative responses could not point to a participant's identity. This encouraged participants to say what they truly thought since they didn't have to be concerned about thier responses being paired to them.

16.2.2 Mitigating the Weaknesses of Online Studies

Online studies are not without weaknesses. Some of these are true limitations but cannot be avoided (see the discussion on limitations in Section 16.3.2). However, many of the perceived weaknesses of online studies can be mitigated, as demonstrated by this case study. For example, the case study researchers mitigated against online participants only providing low-effort responses (such as skipping questions, answering only multiple-choice questions, or providing one-word responses) by motivating participants to provide thoughtful responses through the experiment design and recruitment strategies. With the appropriate participant motivation, remote participation can even lead to higher-quality answers than those of in-person studies, due to anonymity and participants being able to complete responses leisurely (see Section 16.2.3.3).

Perhaps the most obvious disadvantage of online studies is that they are virtualised or simulated approximations and can never truly reflect reality. If the study is not designed well, then it may not be possible to extrapolate the data collected to a real-world in-person scenario. When designing an online study, it is helpful to consider what features are different from the real world, whether the features that are missed or different are important to the topic of research, and how these different features can be substituted or controlled for.

Sound is one element that should be carefully considered for online HRI studies, as it can strongly influence human responses and (depending on the experiment design) can also vary significantly between online versus in-person stimuli. Some researchers choose to mute robot videos to avoid sound being a confounding or uncontrolled factor in online studies. For example, to evaluate the perceived human-likeness and appropriateness of robot gestures generated for speech, an online crowd-sourced experiment conducted on MTurk [Yazdian et al. (2022)] muted the robot videos prior to dubbing them with speech. Similarly, Izui and Venture (2020) muted video stimuli for the online condition due to difficulties in controlling the sound participants experience. However, the decision to remove sound from videos (rather than controlling for sound) should be carefully considered, as in-person experiments with robot motor noise have demonstrated the significant influence of consequential sound on user's impressions towards the robots [Frid et al. (2018), Trovato et al. (2018)]. While it is true that the researcher has no control over a participant's choice of audio setup (headphones, speakers, or sound turned off) in an online experiment, it is still possible to control for the effect of sound within the results by gathering data on what each participant was able to hear and removing any invalid responses where participants were not correctly hearing sound.

For this case study, controls were put in place to ensure people were listening to the sounds as per the design of the experiment. Prior to viewing the videos of robots, participants were given two brief sound tests. The main test checked that a participant's personal headphones or speaker was set to an appropriate volume to hear the sounds. Participants were played three different stimuli (bird sounds, broadspectrum "noise", and silence) and asked after each sound to select what sound they heard. Participants who got more than one of the three questions wrong (2% for this case study) were considered to have failed the test and were thus removed from the dataset prior to any analysis, as there was insufficient evidence that they could have correctly perceived the sound stimulus of the experiment. A secondary test examined which frequencies each participant was capable of hearing across the human audible spectrum, 20 Hz to 20 kHz [Fastl and Zwicker (2007)]. The test contained a series of pure tones, played at non-equal intervals.

Participants were asked to press a button each time they heard any sound, which would then be compared against the known timings of the pure-tone notes. Missing specific frequencies provided evidence of the frequencies individual participants could perceive from the experiment, whether due to their own hearing or the specifics of their sound setup.

Another vector to control for in online HRI experiments is the environment in which the robot is perceived, for example, an office, factory, laboratory, or at home. The environment is usually obvious for in-person experiments but can be open to interpretation for online studies, defaulting to whatever location participants are in when they engage in the experiment. When designing a study to be run online, it is important to correctly position the environment of the study. For the case study, this was achieved by providing filmed footage matching the intended environment and, additionally, asking participants both during the instructions phase and also alongside questions to specifically think about a certain environment when providing responses. While it will not be effective for 100% of participants, researchers can encourage participants to complete the experiment within a given context; for example, the case study recommended participants use a computer (rather than a mobile device). Similarly researchers could suggest that participants complete the experiment from an office, public space, or home location. Some context can also be inferred by recording the device or browser a participant used while completing the study. In the case study, this showed that most people used a laptop or desktop device.

A final way to mitigate potential limitations of online studies is to do subsequent in-person verification of the online results. It is important to plan for this in advance and design the online experiment to match the in-person counterpart as closely as possible, for example, by filming stimuli for the online version in the location in-person studies will be held. Conducting a preliminary online study can also facilitate recruitment for the in-person verification study, as sample size and diversity may be less important for a follow-up verification study.

16.2.3 Effective Recruitment

Recruitment is effective if it fulfils these criteria:

- *Sufficient sample size*: There needs to be a sufficient number of participants to answer the research question (see Section 6.3.1.1).
- *Sample diversity*: The important relevant characteristics of participants (e.g., demographics or familiarity with technology or robots) should vary across the cohort recruited, matching target population variability.
- *Quality, effortful responses*: Participants answer most questions, provide honest responses, and give details for long-form qualitative questions.

The following subsections provide insights into how these recruitment criteria were achieved for the case study experiment.

16.2.3.1 Methods for Recruitment

There are many different methods available to recruit participants to an online experiment. The case study utilised a variety of channels for recruitment to reduce bias and increase overall population diversity. See Section 4.2 for additional techniques to assist participant selection.

Low-effort recruitment strategies include recruitment within the host research organisation and crowd-sourcing options, both of which typically have large "convenience sampling" biases (see Section 4.2). Unfortunately, it is still very common in HRI research for population cohorts to consist almost entirely of university students and/or have a geographic bias to the region near the research facility. For the case study, this bias was mitigated by recruiting from multiple channels and capping the percentage of participants recruited via crowd sourcing. Crowd-sourced participants are typically incentivised to rush, so responses from these sources in the case study were vetted for quality by looking for too many short answers and/or inconsistent Likert responses from any one participant.

Medium-effort recruitment strategies include using the researchers existing personal contacts, research connections, and social media channels. To lower biases from a single channel, the case study recruited from a variety of geographies and global locations when selecting existing contacts and media channels.

Higher-effort recruitment strategies include attending in-person networking meetups or leveraging industry contacts or other related stakeholder groups. This may include industry partners (office or workplace setting), consumers (home setting), hospitals or care facilities (medical setting), professional organisations, community groups, or other venues attended by cohorts of interest. The case study researchers utilised existing robotics and technology company industry contacts as starting points, some of whom were able to distribute recruitment messages via their existing mailing lists or noticeboards. These were effective recruitment channels as they included people connected with the area and problem being researched and thus were motivated to participate and give meaningful responses. This strategy can introduce bias, but was mitigated against in the case study by avoiding using only contacts from the same industries and including a variety of diverse in-person networking events.

Snowballing, where one potential participant additionally invites their own contacts to participate, was used in the case study. Using snowballing can perpetuate bias if initial contacts recruit within a similar cohort. Participants can also be asked if they previously participated in related studies, which can help to check data for biases.

Further recruitment methods are summarised in Table 4.1 of Section 4.3.

16.2.3.2 Measuring the Effectiveness of Recruitment Strategies

It is important for researchers to assess the effectiveness of their recruitment strategies. The easiest measurement for effectiveness of recruitment is sample size; however, this doesn't give any indication of diversity of the recruited cohort. Online experiments facilitate simple ways to collect data on diversity characteristics of participants.

When deciding what to measure, researchers should evaluate which participant characteristics are relevant for their specific research questions and capture diversity in these areas. For the case study, this included demographic criteria (age, gender, geographic location), and HRI-relevant metrics (familiarity with technology, robots, and the specific robot being used). Participant characteristics for the case study cohort can be seen in Figure 16.2. By collecting these attributes, the data can be checked for diversity across these criteria and analysed for any effect of these variables on results. Choice of recruitment channels can also attribute some bias, so researchers may opt to record where participants were recruited from, either by explicitly asking, or using different web links for each recruitment method.

The mode of measuring participant characteristics should also be considered, such as using quantitative or categorical choices versus free-text responses, or opting to auto-collect some of the data. Typically quantitative or categorical metrics are both easier for participants to provide and for researchers to analyse. Many online survey tools provide options to auto-collect some data. Qualtrics was used for the case study to auto-collect the approximate location of participants, which browser they used, when they did the experiment, and how long they spent interacting with various elements. Auto-measuring some data can help to lesson participant burden by making the experiment shorter; however, it is possible that some participants may question data they didn't explicitly provide. There is also the risk that auto-collected data may be inaccurate, for example, if the participant used a VPN (incorrect location) or kept the browser tab open while doing other activities (incorrect times of engagement). If choosing to auto-collect data, researchers should include this in their human ethics applications and provide a statement to participants at the start of the experiment about which data are being auto-collected.

16.2.3.3 Incentivising Quality Responses

Even with well-intentioned participants, it is not unusual to see many qualitative questions left blank or answered in as few words as possible. For the case study experiment, most people filled in a high proportion of the qualitative fields, and often with long sentences or paragraphs. This was achieved through utilising motivation techniques to encourage people to both participate (recruitment) as well as to provide quality responses (experiment design). The experiment was designed to limit trial number and length (reducing participant fatigue), used stimuli which were engaging for participants (short robot videos), and chose question wording to strongly link the problem to the participant. An example question from the case study is: "What sort of sounds would you personally like robots to make? Be as creative or specific as you like. You can include different sounds for different robots or scenarios e.g. at home, with friends, reading a book." The full questionnaire is available on the research project website [Allen et al. (2024a)].

One recruitment method used in the case study to obtain quality responses was to prioritise recruitment channels of stakeholders with an interest in the success of the research; for example, it solves or creates a problem for them. While effective, this should be combined with broader recruitment efforts to avoid biasing results. The case study also utilised snowballing, where one participant invites their own connections to participate. This can be an effective recruitment strategy because participants are often more motivated to provide quality responses if they know the person who asked them to participate. Section 16.2.4 shows data on completeness and length of responses received for the case study experiment that utilised the above techniques.

Although not used in this case study (except for a small number of crowd-sourced participants), a common incentive method is offering participants payment or a chance to win something of value, such as a giftcard. This can be effective for gaining agreement to participate, but it may bias who chooses to participate and doesn't guarantee quality of responses once a participant agrees to participate (see further discussion in Section 4.3.2).

16.2.4 Results from This Case Study

The case study utilised the above-mentioned techniques for experiment design and participant recruitment to successfully recruit 186 participants, in fairly balanced proportions across a range of diverse criteria including geography (16 countries), gender, age groups, and familiarity with robots. As part of the experiment design, sound tests were included to screen for valid, quality participant responses. After removing participants who failed the screening tests, and thus were not experiencing the experiment as intended, the results contained quality data from 182 participants who completed 858 trials. Participant characteristics for the recruited cohort can be seen in Figure 16.2, which displays fairly balanced proportions of participants within each group.

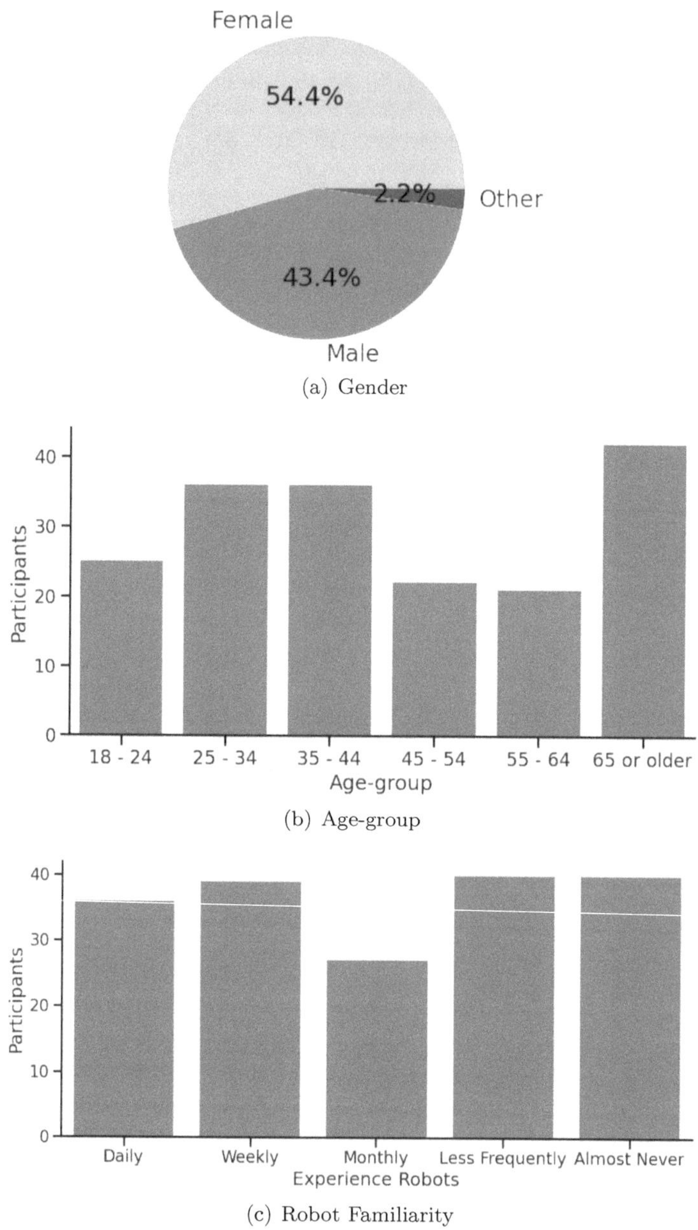

FIGURE 16.2
Balanced diversity of recruited cohort for this case study, as seen on the research project website. Gender, age group, and robot familiarity of participants were just some of the characteristics of participants displaying balanced recruitment across all groups. [From Allen et al. (2024a).]

TABLE 16.1
Qualitative Question Proportions in Mixed-Mode Case Study "Per-Trial" and "Post-Trial" Questionnaires

Condition	**Questionnaire**	
	Per-trial	**Post-trial**
Consequential sound (CS)	15.4% (4 questions)	77.8% (7 questions)
[**control**] No sound (NS)	29.4% (5 questions)	80.0% (4 questions)

A major aim of this case study research was to gather exploratory data on how people want robots in shared spaces to sound to better facilitate co-location with robots. As such, the mixed-mode questionnaires (per-trial and post-trial) contained a number of qualitative free-form questions. The number of qualitative questions asked and the proportion of each questionnaire that was qualitative are shown in Table 16.1, which breaks this down by condition experienced (sound versus control) and questionnaire type. Summary data on the completeness and length of responses received for these questions are shown in Tables 16.2 and 16.3, which provide response rates for *unique responses only* to filter out blank as well as overly simplistic short-form answers such as "No" or "NA". These metrics provide a good proxy for demonstrating the effort put in by participants, and thus anticipated quality and usefulness of responses. Reviewing the content of these responses confirms that the views provided by participants were on-topic and insightful, and achieved the exploratory purpose of this research.

Below are two example responses received for the question, "How do you feel about consequential sounds robots make?"

> The quieter movement sounds are better, they are less distracting and make me feel calmer. The loud noises made me anxious, and I couldn't appreciate the tech in the robot because of it.

> I think there are two parts to how I feel about the sounds 1. Wether I like or dislike the sound 2. How the robot is being used and is the sound appropriate. In situations where a robot is meant to interact with people, the sounds comming from the robbot should not be distracting from the interaction. The closer a robot is mimicking or replacing interaction with a human or an animal, then we should only expect to hear sounds similar to a human or animal. With Pepper, although the gestures convey meaning, the clunky noises distract from that. With Yanshee, I see this as just a fun thing to learn and play with, so I find the servo motor noises pleasing, because I see it more as a toy than a communication tool. If Pepper is just being used for education and fun, then the clunky noises are less of a problem.

With robots like the drone and the four wheel robot, I don't like the sound but I think its appropriate for what I think their use cases are. Ie I don't see them interacting with people so much, but people need to be able to hear them because there would be some danger in being near them.

TABLE 16.2
Qualitative Question Response Rate in All Robots and Response Character Length in Per-Trial Questionnaire

Q_ID	Condition	Topic	Unique Responses, %	Length (mean)	Length (max)
Q10_3	Both	Like/dislike about the robot.	72.7	72	1139
Q12_2	Both	Features of robot that make you happy/unhappy to be in the same space.	77.4	99	596
Q12_3	Both	Comments about this specific video/robot.	23.3	89	531
Q11_3	Both	How you want this robot to sound.	77.2	68	527
Q31_2	NS	Describe what you think the robot would sound like.	95.6	55	455

Note: NS, no sound.

As we can see from these examples, participants provided thoughtful and detailed responses to the questions. This large volume of detailed responses were qualitatively analysed using topic modelling to identify common properties of robot consequential sounds that participants expressed liking, disliking, wanting to hear, or wanting to avoid being produced by robots. Quantified counts of sound-property codes and valence codes showed interesting insights, including some expected results (such as disliking high-pitched and loud sounds), alongside less obvious insights such as wanting informative and audible sounds (over no sound), wanting rhythmic sounds (whether acute or continuous), and more natural sounds (such as cat purrs) in preference to machine-like noises. Further insights specifically related to this case study project can be found on the project website [Allen et al. (2024a)].

TABLE 16.3
Qualitative Question Response Rate in All Robots and Response Character Length in Post-Trial Questionnaire

Q_ID	Condition	Topic	Unique Responses, %	Length (mean)	Length (max)
Q50_1	Both	Why favourite robot?	81.9	86	523
Q50_2	Both	Why least favourite robot?	80.8	87	394
Q53_2	Both	What sounds would you like robots to make?	80.2	177	1033
Q53_3	Both	Any other comments?	47.8	253	1460
Q52_1	CS	Why favourite robot sounds?	71.9	56	288
Q52_2	CS	Why least favourite robot sounds?	73.0	52	261
Q52_3	CS	How do you feel about robot consequential sounds?	65.2	214	1114

Note: CS, consequential sound.

16.2.5 Benefiting from Recruitment Beyond the Initial Experiment

The recruitment efforts for one experiment can be useful beyond that single experiment, for example, by enabling recruitment for a follow-up in-person version of the same experiment, allowing dissemination of the results of the study, or providing a list of interested participants for doing future studies in a related area.

For the case study, once the experiment had concluded, the experiment landing page was repurposed for future recruitment by replacing it with a

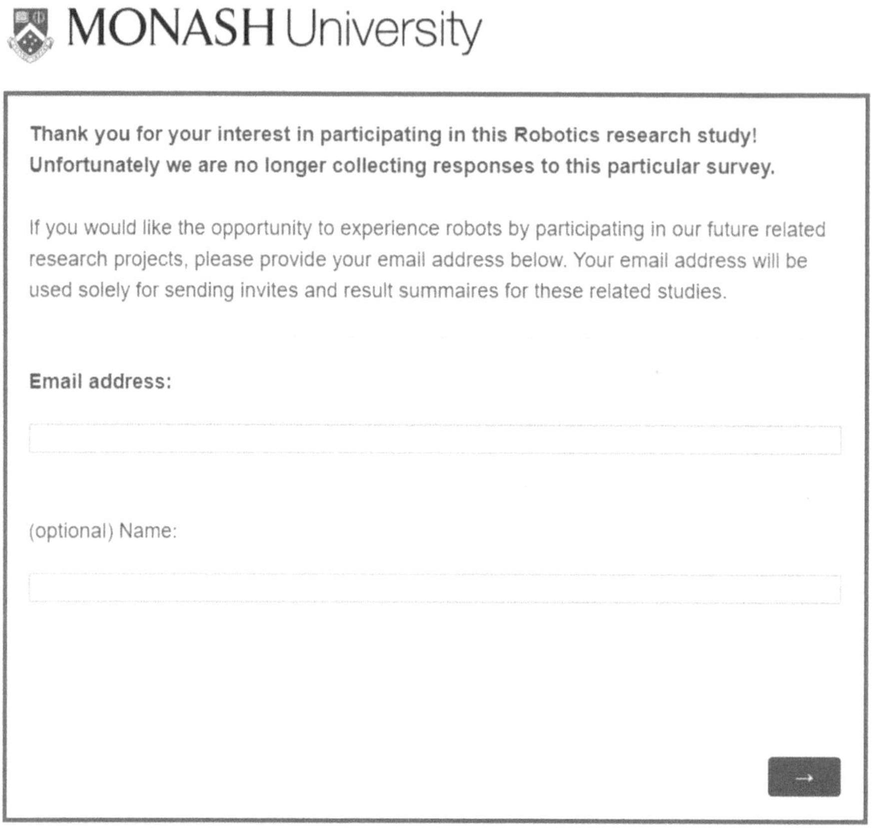

FIGURE 16.3
Example landing page making use of existing recruitment from previous experiments.

page stating that that specific experiment was over, and offering people to participate in future related studies, using a small form to collect an email address and optional name (see Figure 16.3 for an example). When running the next related study, this landing page can again be replaced with a message and hyperlink to the current study. Existing recruitment entry pathways (such as posters and social media posts made during prior active recruitment) directed people to this landing page and led to passively capturing 21 interested participants for new studies across several months.

Follow-up experiments, such as in-person verification or the next stage of related research, can sometimes benefit from having some repeat participants. Recruitment channels from an earlier experiment can be utilised to save effort in future recruitment by collecting contact details during the earlier experiment. Depending on human ethics requirements, it can be good to do this anonymously by having a separate one-page contact details form (such as a small Google form) that is not part of the same survey. This form can be linked to (or embedded) on the final "thank you" page of the main survey, thus providing participants with an easy way to leave their contact details while maintaining anonymity from their already provided results.

It is important if following these recruitment strategies to ensure that they have been reviewed and approved by the appropriate human research ethics committee, for example, permission to collect contact details (email address and optionally name), and recruitment via social channels and immediate contacts of researchers. Likewise, it is important to provide clear and accurate information to participants regarding what contact details may be used for (e.g., only to contact regarding invitations to participate in, or results of experiments run by one of the researchers of the original experiment).

16.3 Discussion

In this section, we summarises the benefits and limitations of online HRI studies, as illustrated by the case study on robot's consequential sounds.

16.3.1 Benefits of the Online Format

The key benefits obtained included:

- Large *sample size* obtained by utilising the lower costs of recruitment and participation for online studies, leading to 182 participants completing 858 trials.
- Strong *diversity of cohort* from utilising a variety of recruitment strategies to obtain large geographic and demographic reach, with results showing fairly even recruitment on key diversity criteria across the cohort (Figure 16.2).
- *Online-only benefits* were taken advantage of such as the ability to have identical stimuli for participants and better controls such as implementing a no-sound condition.
- High volume of *quality responses* obtained through the benefits of online anonymity and design and recruitment strategies to incentivise participant engagement (particularly useful for exploratory research and underresearched domains).
- *Future in-person studies* can be better informed through running this online study.

16.3.2 Limitations

Despite the benefits of conducting online studies as just outlined, there are several limitations illustrated by the case study:

- *Inability to reach certain demographics*: It can be difficult to reach people in certain countries or rural areas, those with limited Internet access, or those low information technology literacy. Case study participants were mostly from larger cities across 16 developed countries.
- *Limited knowledge or control of environment of participants*: Environments can be quiet versus busy, and access to the experiment can be by mobile versus computer. The case study recommended participants use a laptop or desktop computer; however, responses showed several people still participated via a smartphone.
- *Not truly reflecting the reality of physical co-location with the robot.* The case study made every attempt to capture a realistic visual environment and record full-spectrum sounds but could not replicate the 360 degree in-person ambient experience.
- *May not be as effective for longitudinal studies*: It can be hard to re-engage people online, even if they committed to the entire study the first time.
- *Not good for experiments where the participant requires a lot of training*: The experimenter will not be there to provide help, so the method of participation and successfully completing trials needs to be intuitive and easy. For the case study, only a small minority of people expressed issues with using the online platform due to lack of familiarity or limited information technology literacy.

16.4 Summary

This chapter utilised a case study HRI experiment to demonstrate some of the key benefits of online studies, and techniques researchers can use to leverage these. The experiment investigated how consequential sounds produced by robots affect human perception of robots by showing participants videos of five different robots before surveying their opinions on these robots.

Leveraging Online Studies for HRI Research

- Online studies offer benefits including:
  - The ability to provide identical stimuli to participants.
  - Increased participant diversity.
  - Larger sample sizes, stemming from reduced costs and ease of participation.
  - Quality responses may be encouraged by anonymity and appropriate incentivisation; this works in conjunction with large and diverse samples to create strong datasets.
- Limitations of online studies include not reflecting reality as faithfully, inability to reach certain demographics, and not being well suited to high-training or longitudinal studies. Some limitations can be removed or mitigated through controls.
- Tips for conducting online studies:
  - Carefully choose recruitment channels and include a variety of channels.
  - Design the experiment to easily measure the effectiveness of recruitment and diversity of cohorts.
  - Design online studies to incentivise quality responses.
  - Set up recruitment efforts to continue providing benefit for future research.

17

Measuring Interaction with Multimodal Data

Zhongxiang Chen

Human-Robot Interaction comprises different aspects: physical, social, and cognitive. During such diverse interactions, a variety of physiological, psychological, and behavioural measures may be of interest, giving rise to the need for multimodal measurements. In this chapter, we provide a case study describing a multimodal data-collection experiment focusing on physical HRI, investigating human gait adaptation to a wearable passive exoskeleton.

17.1 Multimodal Measurement of Human Data

As was discussed in Section 5.2, human measurement during HRI can be collected from various sensor and measurement modalities, such as video recordings of task performance, physiological measures of human responses and behaviours, and questionnaires on psychometric scales and perceptions. While the inclusion of multiple modalities provides the capability for a richer, more nuanced assessment of the interaction, multimodal measurement also entails additional experimental complexity and effort, requiring careful experimental design, data collection, and analysis procedures.

17.2 Case Study: Human Adaptation to Kinematic Constraints

As an example of a multimodal HRI experiment, we discuss a study focusing on human gait analysis to support assistive exoskeleton and prosthesis

DOI: 10.1201/9781032693682-20

design. Gait analysis is crucial for understanding human locomotion and generating human-like gait for exoskeletons and humanoid robots and has attracted attention from various fields, including biomechanics, rehabilitation, and robotics [Kulić et al. (2016)]. Studies have shown that humans optimise their gait and adapt it to various factors such as speed, external environment, and age. To understand human biomechanics, disease and injury recovery processes, and how to design effective assistive and support systems such as exoskeletons, it is crucial to understand both nominal gait patterns as well as gait adaptations. However, research on gait analysis and gait synthesis faces a number of limitations: Accurate data collection is resource-intensive, and the data collected are often not publicly shared. Additionally, when analysing gait adaptations, many studies focus on external constraints (e.g., walking speed) rather than internal factors (e.g., muscle strength). Furthermore, comparisons across conditions usually rely on data averaged over a participant group instead of examining individual-specific adaptations, leaving a gap in understanding person-specific biomechanical adjustments for enabling personalised HRI [Abbasi-Bafghi et al. (2012), Lim et al. (2022)].

17.2.1 Study Aims

The aim of this case study was to address the research gap in studying individual adaptations to constraints. Gait may be modified by external constraints, such as obstacles and other moving objects, and internal constraints, such as muscle weakness and limited range of motion. This case study investigated the impacts of external constraints through controlling the walking speed and step frequencies of the participants and simulating an internal constraint of limited left knee flexion by applying a passive knee orthosis. The walking speeds were controlled by setting the treadmill speed (0.4, 0.8, and 1.1 m/s). Step frequencies (10%–20% of preferred frequency) were introduced by asking participants to follow a metronome audio cue. The total number of testing conditions, combining the three walking speeds, five step frequencies, and with or without left knee constraint, was $3 \times 5 \times 2 = 30$. To accurately record the resulting gait patterns, a multimodal dataset was collected, consisting of four types of information: (1) kinematic data of reflective markers attached to body landmarks, collected by the Vicon camera motion capture system, (2) ground reaction forces generated by the contact between the left and right foot, collected by the dual-belt treadmill, (3) metabolic cost, collected by the PNOE metabolic cost analyser, and (4) muscle activations of selected leg muscles, collected by the Delsys Trigno system, as shown in Figure 17.1. The multimodal data collected has been published to provide a publicly available dataset and enable research on person-centric gait analysis [Baček et al. (2024)]. The study protocols were reviewed and approved by the university's human research ethics committee.

FIGURE 17.1
Experiment setup for multimodal data collection. [From Baček et al. (2024).]

17.2.2 Recruitment Process

Recruitment was advertised on the Human Robotics Research Group's official website at The University of Melbourne, Australia. The objectives, procedures, and duration were outlined on the advertisement, and contact information was publicly available for those interested in participating. The participants had to meet the following inclusion criteria to be eligible: (a) no known lower limb impairments, (b) 18–60 years of age, (c) 1.55–1.90 meters tall, and (d) body weight 50–90 kg. Participants were excluded from the study if any of the following conditions applied: (a) significant recent trauma or orthopaedic procedures on the lower limbs, (b) diagnosed gait abnormalities or walking difficulties, (c) cognitive or communicative issues affecting instruction comprehension, or (d) other pain conditions.

17.2.3 Experimental Protocol

The experiment was organised into three sessions. A participant came to the CAREN lab at the University of Melbourne on different days for each session. Prior to participant arrival, the 10 Vicon cameras were calibrated with an Active Wand, and the Delsys Trigno EMG sensors and PNOE metabolic analyser were fully charged. Upon participant arrival, 26 spherical reflective markers were placed at prominent bone landmarks on the skin and to differentiate right and left body segments in the Vicon system. The participant's skin was disinfected by alcohol swabs, and excess hairs were removed before placing the reflective markers and EMG sensors. The EMG sensors were placed along muscle fibers over the tibialis anterior (TA), gastrocnemius lateralis (GL) and medialis (GM), vastus lateralis (VL) and rectus femoris (RF), biceps femoris (BF) and semitendinosus (ST), and gluteus maximus (GMax). The forceplates under the dual-belt treadmill and Delsys Trigno system were connected to the Vicon system for data sync purposes. The PNOE metabolic analyser required calibration every 30 minutes, and the data were processed on the server asynchronously. Thus the metabolic cost measurements were synced post data collection.

The first session, the preparatory session, allowed for the participants to become familiar with the equipment settings and for the investigators to collect the preferred stepping frequencies of the participants at six different conditions. The six conditions consisted of three walking speeds: 0.4, 0.8, and 1.1 m/s, and two kinematic constraint conditions: with and without a passive knee orthosis limiting knee flexion. This session took approximately 2 hours with 80 minutes of walking. It allowed the participants to become familiar with the sensors and the task, which reduced the influence of the data-collection equipment on their gait patterns. Participant training and preparation was previously discussed in Section 4.5. The two following data-collection sessions had the participants following instructions to walk at 30 conditions in total. The six conditions mentioned for the preparatory session served as the baseline of the 30 testing conditions. The participants were asked to follow audible cues from a metronome to adjust their stepping frequencies at those six conditions. There were five stepping frequencies: –10%, –5%, preferred, 10%, and 20%.

17.2.4 Results

Human gait is commonly assumed to be optimised according to implicit objectives and preferences [Alexander (1984)]. The large number of joints of the human body permit numerous ways for locomotion. A different gait pattern may be selected based on factors such as travelling speed, altitude elevation, and surface stiffness. Despite the large number of possible locomotion strategies, studies have shown that the average usual preferred walking speed at usual pace is around 1.31 m/s for healthy adults [Murtagh et al. (2020)].

Twenty-one participants enrolled in the study, with two participants withdrawing after the first data-collection session. Therefore, the analysis and figures presented in this chapter exclude the two withdrawn participants.

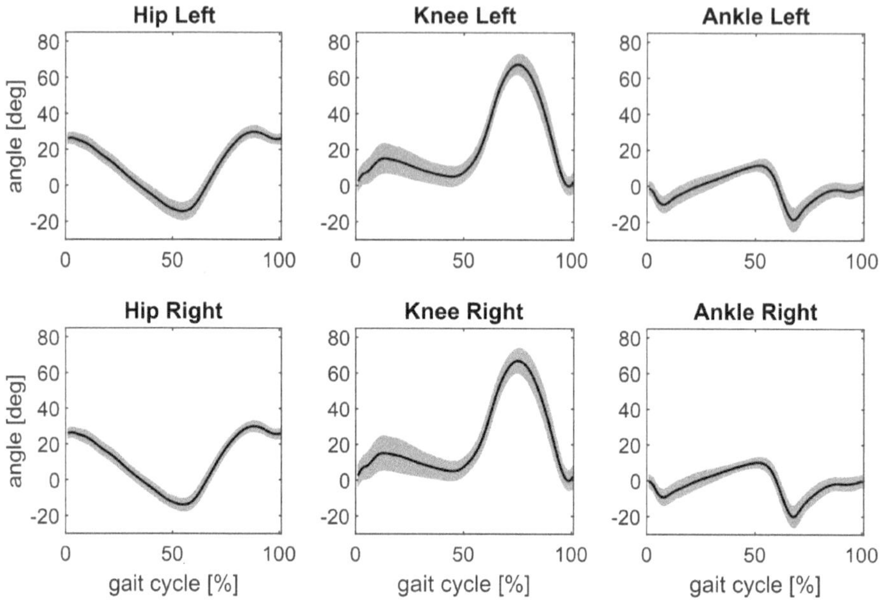

FIGURE 17.2
Joint trajectories of all participants walking at 1.1 m/s at their preferred cadence without the passive orthosis. The solid line is the average joint trajectory, and the grey area shows the average ± one standard deviation.

The publicly available dataset [Baček et al. (2024)] contains all the raw data and scripts for basic data processing and visualisation. The *Segment Data* section in Baček et al. (2024) contains joint trajectories of sagittal joints only. Three walking conditions were selected, and the *segmented data* of the 19 subjects were aggregated and presented. Figure 17.2 shows the joint angle trajectories of the hip, knee, and ankle joints in the sagittal plane when subjects were walking at 1.1 m/s at their preferred cadence without induced constraints. Figures 17.3 and 17.4 show the joint trajectories when participants were walking at 0.4 m/s at their preferred cadence without and with the passive orthosis, respectively. In all three plots, the black line indicates the mean joint trajectory and the grey area encompasses the range one standard deviation away from the average. The average preferred walking speed without the orthosis was found to be 1.1 ± 0.12 m/s for the 19 subjects.

As can be seen from Figures 17.2 through 17.4, when participants were forced to walk more slowly than their preferred speed, the variances were larger. Furthermore, the variances in joint trajectories on the unconstrained side, the right side, increased further when the passive orthosis was applied to the left knee. The increase in variance corresponds to the different adaptation strategies observed from the participants.

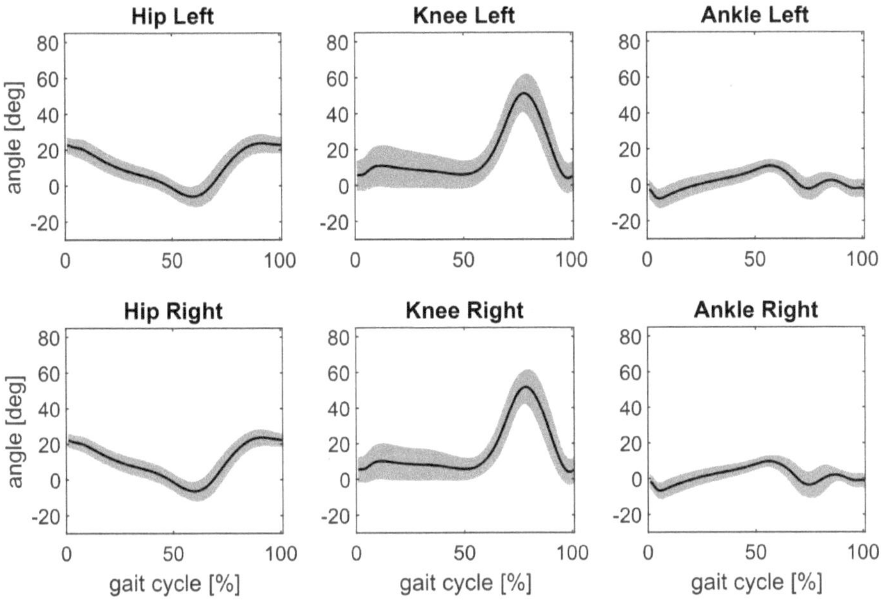

FIGURE 17.3
Joint trajectories of all participants walking at 0.4 m/s at their preferred cadence without the passive orthosis. The solid line is the average joint trajectory, and the grey area shows the average $\pm$ one standard deviation.

The raw data from the Vicon system were further analysed with OpenSim [Delp et al. (2007)] to investigate these individual adaptation strategies in more detail. A lower extremity skeletal model was firstly scaled with marker locations during the static calibration pose. The scaled model was then used to track the marker locations during walking through the built-in *Inverse Kinematic* function in OpenSim. A number of compensation behaviours were identified from the kinematic data. For example, participant 9 exhibited circumduction, which is "greater-than-normal frontal thigh angle during midswing of the affected leg" [Kerrigan et al. (2000)], while participant 11 adopted step-to-walking, a gait where the "toe of the affected side does not pass the toe of the non-affected side in the stance phase" [Inoue et al. (2022)], when walking with the passive orthosis on the left knee. Figure 17.5 shows a single gait cycle for both participants 9 and 11 walking at 1.1 m/s at their preferred cadence without the orthosis. Figure 17.6 shows a single gait cycle for these two participants walking at 0.4 m/s at their preferred cadence with the passive orthosis. Due to their different adaptation strategies, a clear increase in the left hip rotation was observed from participant 9 to achieve circumduction. On the other hand, a smaller right hip extension was observed in the data for participant 11, which was caused by the step-to-walking behaviour.

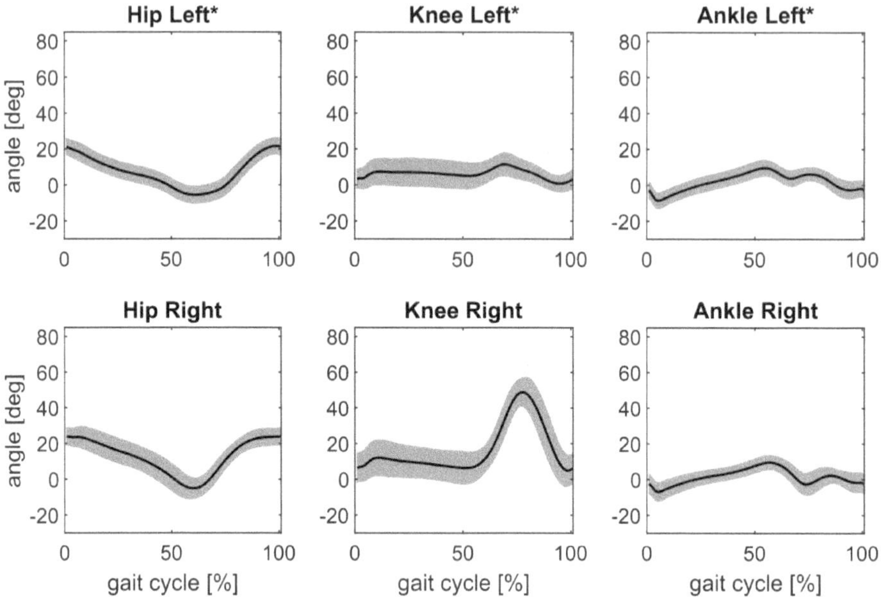

FIGURE 17.4
Joint trajectories of all participants walking at 0.4 m/s at their preferred cadence with the passive orthosis. The affected side (left knee) is indicated with $*$. The solid line is the average joint trajectory, and the grey area shows the average $\pm$ one standard deviation.

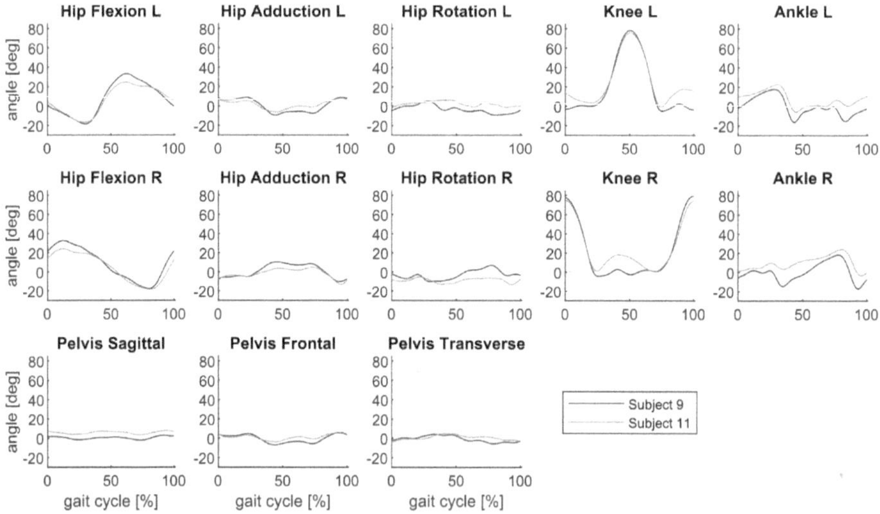

FIGURE 17.5
One gait cycle of the joint trajectories of participants 9 and 11 walking at 1.1 m/s at their preferred cadence without the passive orthosis.

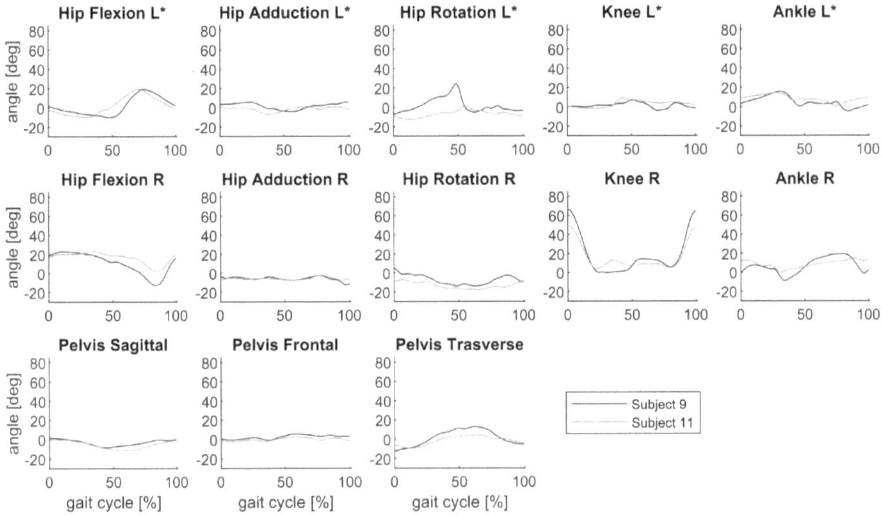

FIGURE 17.6
One gait cycle of the joint trajectories of participants 9 and 11 walking at 0.4 m/s at their preferred cadence with the passive orthosis. The trajectories on the orthosis side are indicated with $*$. Participant 9 adapted to the orthosis by circumducting the hip, while participant 11 adopted a step-to-walking strategy.

17.3 Discussion

In this section, we discuss considerations of multimodal data collection in HRI research, as reflected by the case study.

17.3.1 Number of Measurement Modes

Four types of data were collected during the study: the marker trajectories of reflective markers, left and right foot ground reaction forces, metabolic cost, and EMG amplitude of muscles in both legs. Measuring multiple types of data enabled the researchers to study the details of gait mechanics from various perspectives, including kinematic, dynamic, energetics, and muscle recruitment. The multimodal data provided a holistic view of how different systems interact during movement. This holistic approach permitted the study of the intricate details of gait adaptations, the impact of interventions, and the resulting metabolic cost. As was discussed in Section 5.3, combining multiple measurements allows for a comprehensive understanding of human behaviours and interaction, as well as contributes to increased validity of the study design.

However, there are many challenges associated with multimodal data collection. The technology and equipment needed to collect and synchronise these data are often expensive and require extensive setup and frequent calibration to ensure precision and alignment. This often constrains the study to a specialised laboratory setting with limited accessibility. In the case study, the equipment required thorough cleaning and disinfecting to ensure the safety of the participants, as reflective markers and EMG sensors were placed directly on skin, while the PNOE metabolic analysed measures all air flow through the respiratory tract. Training was needed to ensure all investigators correctly used the equipment. Additionally, analysing multiple data streams simultaneously demands computational resources, sophisticated software, and expert knowledge, which can be both time-consuming and prone to error. These constraints led to the multimodal data collection often having a smaller sample size compared to crowd-sourced data collection (see Section 3.2.1) with only one or two metrics.

In summary, while the comprehensive insights gained from multidimensional data are valuable, they come with substantial demands in terms of resources and analytical complexity. Thus, it is recommended that researchers weigh the benefits and risks of different data-collection modalities in relation to their research aims and available resources.

17.3.2 Number of Test Conditions

Testing a large number of conditions in human gait experiments offers a broad perspective on how various factors influence movement, but it also presents several challenges. The participants of this study were both the control and test group; that is, data on both the and constrained gaits were collected from the same subjects, allowing direct comparison between the baseline and behaviour following the intervention. The 30 combinations of walking speed, stepping frequencies, and left knee kinematic constraint revealed how these factors individually and interactively affected gait biomechanics. This comprehensive approach enabled insights of the full spectrum of gait adaptations to be gained. Furthermore, participants were instructed to walk for 5 minutes at each test condition. This was chosen to enable both the transient and steady-state responses to the change of test conditions to be observed. However, there were several drawbacks to testing many conditions.

Firstly, the data-recording process was prolonged and the resulting analysis became more complex. The main data-collection phase of the study was split into two sessions, each lasting around 3.5 hours and requiring participants to walk for about 80 minutes. This was very physically demanding, thus limiting the selection of participants to those who were relatively fit and young. The other option was to split the data collection into more sessions. However, that would have made the overhead of the equipment preparation more significant and scheduling resources more challenging. In addition, four types of data were collected through different channels. More test conditions leads to a higher risk of experiencing issues, such as losing reflective markers, depletion of batteries, and recalibration requirements. Thus, planning for possible failures and contingency is required to continuously collect high-quality data, as was discussed in Section 3.2.5. The sheer number of conditions can also dilute the focus of the study, making it harder to draw clear conclusions or identify the most significant factors affecting gait. Furthermore, the effect of fatigue should be accounted for in the analysis of the gait adaptation and comparison between different conditions.

Therefore, while exploring a wide range of conditions can provide valuable insights into human behaviours influenced by various factors, it demands careful planning and experimental protocol design, advanced analytical techniques, and a thoughtful approach to data interpretation to manage the associated complexities effectively.

17.3.3 Number of Participants

Increasing the sample size enhanced the robustness and generalisability of the study's findings. The larger participant pool was able to capture a wider range of gait variations due to factors such as age, sex, body type, and health conditions, thereby providing a more comprehensive understanding of how gait patterns vary across different populations. This could lead to research findings more relevant for diverse groups. Furthermore, a large sample size helps to improve the statistical power of a study and allows for more nuanced analyses of subgroup differences. However, there are considerable challenges associated with recruiting a large number of participants, especially in multimodal data collection using specialised equipment, as was shown in this example study. Managing a large sample requires extensive logistical coordination, including recruitment, scheduling, and data collection, which can be resource-intensive and time-consuming. Ensuring consistency in how each participant is tested and maintaining data quality across a large group can be challenging, potentially introducing variability that may confound results. In addition, issues such as participant variability and non-compliance can further complicate data interpretation and may require additional statistical adjustments to account for these factors.

Consequently, while a larger number of participants can enhance the validity and generalisability of HRI research, it demands careful planning, resource allocation, and analytical rigour to effectively manage the associated challenges. When collecting multimodal data from a large number of participants is difficult to achieve due to accessibility and resource constraints, researchers may consider alternative analytical approaches to null hypothesis significance testing, as was discussed in Chapter 10.

17.4 Summary

In this chapter, we reviewed the study protocol and experimental setup used in measuring multimodal data from young healthy adults with induced kinematic constraints. The data collected permitted the researchers to characterise different adaptation strategies of individuals under various speed, cadence, and constraint conditions. The multimodal data-collection process provided a rich dataset, enabling a detailed analysis, but introduced several complexities and challenges.

Collecting a Multimodal Dataset in HRI Research

- Multimodal data provides a comprehensive view of interactions, capturing nuances that a single modality might miss.
- Combining data from different sources helps HRI researchers develop a holistic understanding of human behaviours, intentions, and emotions, and thus contributes to more accurate human models and more effective interactions.
- Integrating and processing data from multiple modalities is technically complex, requiring sophisticated algorithms and significant computational resources.
- Multimodal data collection may limit the participant sample size and diversity, due to the additional experimental and logistical challenges associated with multimodal data collection.

18

Challenges and Future Directions of Experiments in HRI

This book describes a systematic approach to designing and executing an HRI study over its life cycle. The main steps and applicable guidelines at each stage are provided, together with an interactive tool (`https://doi.org/10.26180/26507314`) to serve as a guide and checklist and facilitate study development and monitoring. The interactive tool provides a framework for systematically considering the study design, documenting the design choices, and sharing the developed designs between researchers. For each step in the life cycle, we also provide illustrative examples drawn from the HRI literature.

We provide practical guidelines and illustrative examples to support HRI researchers as they incorporate human-centred methodologies into their studies. A community effort is required for the field to develop and adhere to a framework of ethical HRI, focus on innovative and interdisciplinary discoveries, and address the technical and resource challenges in contextualised HRI evaluation. Thus, our goal is to encourage dialogue within the HRI community and across other related fields to address the challenges of ethical HRI.

This book provides an expanded but non-exhaustive set of HRI methodologies drawn from a human-centred and interdisciplinary perspective. While we have focused on the HRI (and to some extent the HCI) literature, appropriate methodologies continue to be adapted from complementary fields, for example, psychology, neuroscience, and biomechanics. The HRI field is young and rapidly evolving, and adaptation from other fields, where beneficial and highly relevant for HRI, should be considered in experimental design. No doubt, new methodologies and technologies will continue to emerge in related literature. Such new methodologies and technologies may demonstrate that previous theories or ways of measuring phenomena are no longer accepted or conventional. Researchers are encouraged to continue to monitor the HRI and related literature to remain up to date with advancements in methodology.

One example where best practice is likely to change in the future is in study registration. While the importance of identification and registration of hypotheses in confirmatory studies has been acknowledged by the HRI

DOI: 10.1201/9781032693682-21

research community, it is not yet a common practice in the literature. This may become a requirement soon with more publication venues recognising that this approach adds rigour to HRI studies, similar to in behavioural and clinical research studies. In addition, there is an increasing community effort towards sharing open-source code and data to allow for reproduction of HRI studies and easy bench-marking of shared tasks, such as the Habitat simulation environment for home robots [Puig et al. (2023)]. Another area of growth and change is the use of online studies and interaction tools, particularly given the large shift to online interaction during the COVID-19 pandemic.

Our discussions focused on HRI methodologies in the research sector. There are significant additional factors to consider for commercial and broad-scale deployment, similar to considerations in the commercial application of a new medical intervention after Phase 0 to IV clinical trials. Additional factors to consider in HRI applications include public policy considerations and privacy and responsible use of data. Similarly, Chapter 8 discussed recommendations for ethical HRI research in terms of its conceptualisation, design, and evaluation. Ethical considerations in commercialisation and HRI applications, requiring the collaborative effort of researchers, designers, communities, industry, and policymakers to identify the responsible entities of the technology and to ensure ethical and sustainable practices [Boden et al. (2017), Müller (2020), Mintrom et al. (2022)], were not addressed as this book focuses on HRI experimental research.

18.1 Current Challenges and Outlook

We identify the key challenges in HRI experimental research in this section by summarising the discussion in this book, as well as reflecting on existing guidelines of HRI research methodologies, which were reviewed in Sections 1.3 and 9.2.

18.1.1 Ethical HRI Research

Adopting human-centred design in HRI is essential to achieve intended outcomes that benefit users. We adapted the IEEE standard 7000-2021 [IEEE (2021)] to provide suggestions on how to conceptualise HRI research questions in relation to stakeholders and use context, how to identify ethical values and generate requirements to be validated, and how to identify and mitigate risks and encourage transparency, as was shown in Figure 8.1 of Chapter 8. As research on human, AI, and robot ethics progresses, new ethical standards and guidelines are being developed, including the recent IEC and ISO report on the responsible adoption of AI ISO/IEC TR 24368:2022 [ISO (2022)]. New metrics, such as fairness [Bellamy et al. (2018)] and social acceptance [Heerink et al. (2009)], can also be utilised to evaluate outcomes of HRI systems in alignment with ethical values and requirements of the stakeholders. Thus, methodologies and guidelines for ethical HRI are expected to grow and evolve.

Engaging with a broader range of stakeholders beyond the immediate users allows for the conceptualisation of HRI research within specific social and cultural contexts, thus facilitating the identification of relevant values and ethical requirements of HRI systems. Value identification and validation is an active research topic in design ethics (see Section 8.3), as there can be conflicting values between stakeholder groups or shifting values over the design and/or deployment cycles of HRI systems. The value-sensitive design framework emerging in HCI [Friedman et al. (2017)] can be adopted in HRI studies to validate that the experimental design contributes to the progression of ethical HRI systems that align with identified stakeholder values. For HRI researchers, reporting the value identification and inclusion and exclusion process can provide context to the ethical concerns considered during the system or study design. Another emerging research direction that empowers end-users is the development of new robotic systems that assist users in identifying and avoiding over-reliance on, misuses of, or manipulations by, robots or other technologies [Oviatt (2021)].

18.1.2 Foster Interdisciplinary Collaboration

A growing number of HRI studies are investigating human-centred design approaches and are increasingly adopting methodologies from related domains, such as user experience design or social anthropology [e.g., Prati et al. (2021)]. In addition, interdisciplinary exploratory and hypothesis-generating studies can facilitate the investigation of the complex mental and physical interaction dynamics during HRI [e.g., Rueben et al. (2021)]. Moreover, there is a need for new metrics that reflect users' perceptions and cognition during interaction, as well as novel measurements that assess the contextualised task objectives and ethical outcomes of HRI systems, such as classification accuracy metrics addressing potential biases in the data or the model [Bellamy et al. (2018)].

HRI studies benefit from adapting theories from psychology and social sciences to develop hypotheses, constructs, and measures and incorporating innovations for measurement, data processing and data analysis from statistics, machine learning, user experience, and design. However, it can be difficult for HRI researchers to identify suitable theories and methodologies in other fields as consumers of these studies. Thus, high-quality reviews by domain experts in collaboration with HRI researchers are required for the field to access and understand the state of the art of these related fields. Further, as interdisciplinary research often faces difficulties in identifying suitable venues for publishing and funding, support from publications, reviewers, editors, and funding agencies can accelerate innovative interdisciplinary studies by setting up special calls and joint conferences.

Recent advancements in generative artificial intelligence (GenAI) and related applications [Achiam et al. (2023)] have provided new opportunities for HRI. For example, Large Language Models (LLMs) have been used to increase a social robot's dialogue capabilities to open-domain user inputs [Kim et al. (2024)]. Further, foundation models for robotics, especially vision-language-action models such as the Robot Transformer (RT-X) model [O'Neill et al. (2024)], have shown potential to greatly improve robots' capabilities to understand natural language commands and communicate task progress in an explainable manner [Liu et al. (2023)]. However, these pre-trained models can incur a high computational cost not suitable for a robot's available hardware or in-field deployment. In addition, simple prompt engineering may not be sufficient for adopting LLMs in HRI: Fine-tuning with new data or development of new HRI-specific models may be necessary to achieve task success. Moreover, openly accessible LLMs are designed to be general-purpose, while personalised interaction is essential to many HRI scenarios. Furthermore, as discussed by Williams et al. (2024), the use of LLMs in HRI may raise ethical concerns, such as a lack of trustworthiness caused by the inconsistency and hallucination issues in current models. There is a need for new foundational AI models for HRI with increased reliability and transparency.

18.1.3 Increase Contextual Validity

To incorporate contexts in an HRI study, human-centred and collaborative design approaches that involve the stakeholders early on [Duque et al. (2019), Alves-Oliveira et al. (2021)] are effective. Mixed methods including experimental, quantitative, qualitative, and phenomenological research, are helpful to embed the evaluation of an HRI system in its intended social and cultural contexts [Seibt et al. (2020), Veling and McGinn (2021)]. As discussed previously, interdisciplinary collaborations are necessary for understanding and investigating contextual factors in HRI.

Improved practices in study setup that increasingly resemble the application contexts, especially with increasing design maturity and interpretation of experimental results within the study context and assumptions, advance understanding of HRI in-situ, avoiding costly failures [Hoffman (2019a)] and increasing the impact of the field. In-the-wild HRI experiments are challenging to conduct. Thus, publications and reviewers are encouraged to be open to exploratory studies that are innovative but may have limited statistical power. Such studies are important scaffolds towards more mature designs and long-term field trials. This is similar to practices in other fields, such as clinical trials in medicine, where a phased approach is taken: Phase 0 studies are exploratory; Phase I studies evaluate the safety of a new intervention with less than 100 participants; Phase II studies evaluate the effectiveness of the intervention with hundreds of participants; Phase III studies compare the new intervention with other alternatives on hundreds to thousands of participants; and Phase IV studies are long-term evaluation [Australian Clinical Trials (2015)].

18.1.4 Improve Experimental Rigour and Transparency

To improve the experimental rigour and transparency in empirical HRI studies, there is an increasing interest in pre-registering HRI study hypotheses [Hoffman and Zhao (2020)]. For experiment design and evaluation, it has become more common to collect multiple evaluation metrics under more naturalistic settings that reflect the usage context [Bethel and Murphy (2010)]. Regarding HRI study participants, more efforts have been put into recruiting participants who are representative of the potential user population of the HRI system being analysed [Bartneck et al. (2020b)]. Regarding data collection and analysis, recent HRI studies have focused on strengthening the validity and generalisability of data analysis [Schrum et al. (2020)]. While efforts to improve the experimental rigour in empirical HRI studies are critical to advancement in the field, it is worth noting that some of these recommendations can be difficult to implement. For a single HRI study, it may be unnecessary to address all the recommendations and be valuable to focus on the most relevant ones. For instance, pre-registration of hypotheses is important for confirmatory HRI studies but may be less suited for some exploratory HRI studies.

A detailed description of the HRI study process, including any methodology or data-processing errors, should be reported in publications to avoid misleading researchers on experiment results and contributing to the replication crisis as seen in other fields. Publications, reviewers, and editors can help establish and maintain transparency standards. There is an increasing community effort towards publishing open-source code and data to support reproduction of HRI studies and easy bench-marking of shared tasks, such as the Smart City Robotics Challenge [Wang et al. (2020)]. While reproducibility is important to aid collective efforts to advance the field, it is worth noting the difficulties in achieving this due to limited availability of tools and resources, which requires contribution from within and outside of the HRI research community. Transparency in communicating an HRI system's capabilities and limits is also important for users to establish a realistic understanding of how it functions, which reduces the risk of over-reliance on the system or falling victim to predatory use of the technology [Parasuraman and Manzey (2010)]. A transparent communication between HRI researchers and stakeholders can be achieved via participatory design approaches, as was discussed in Section 11.1.7, or by developing explainable computational models and incorporating the explanations with the interactive robot behaviours [e.g., Tabrez and Hayes (2019)].

18.2 Future Directions

Considering the discussion in this book, and the majority of HRI studies to date as reviewed in previous surveys [Bethel et al. (2020b), Winkle et al. (2023), Spitale et al. (2024)], we highlight the following recommendations to improve future research:

Increase awareness and use of approaches in related fields: The HRI community can benefit from closely following and incorporating findings in related fields to improve HRI study methodology. This includes adapting theories from psychology and social sciences to develop hypotheses, constructs, and measures and incorporating innovations for measurement, data processing, and data analysis from engineering, statistics, and machine learning.

Increase use of datasets, bench-marks and foundation models: HRI research can be accelerated by developing robot functions using bench-marking datasets and environments, especially those with high ecological validity. Further, foundation models can increase the autonomy and reliability of HRI systems. However, there is a gap in foundation models purposefully developed for interactive robots and concerns regarding the computational costs of large models.

Increase transparency: A detailed description of the study process, including any methodology or data-processing errors, should be reported to avoid misleading researchers on experiment results, which can further contribute to the replication crisis as seen in other fields. Authors are encouraged to consider pre-registering their hypotheses and open-source the code and data used in their work. Publications, reviewers, and editors can also play an important role in establishing and maintaining transparency standards.

Increased rigour: Authors should strive towards improved practices in recruitment, aiming for larger cohorts with appropriate demographics and systematic sampling, blinding, and allocation methodology. Rigorous statistical methods for data analysis need to be employed.

Adoption of best practices and increased rigour in the design and deployment of HRI studies will inevitably lead to better reproducibility of findings and ensure that findings are relevant in application deployments, avoiding costly failures [Hoffman (2019a)] and increasing the impact of the field.

18.3 Summary

HRI is a quickly developing research area with many challenges and opportunities. Moving forward, ethical research approaches, interdisciplinary collaboration, real-world studies, and improved experimental rigour will serve the field well as it advances. We conclude with a set of general recommendations for human-centred HRI experimental methodologies.

Human-Centred Methodological Guidelines for HRI Research

1. **Goals**: For confirmatory studies, consider pre-registration, and in doing so, document the behavioural or interaction theories that the hypothesis is based on (if applicable), and name the key constructs and factors that will be tested for regarding the human, the robot, and the interaction. For exploratory studies, state the research questions to be investigated and present the emerging observations in an unbiased manner.

2. **Contexts**: Conduct HRI studies within suitable contexts if possible, including but not limited to, social and cultural backgrounds, individual user's personal context or the intended user population's shared context, spatial and situational contexts of the location, risk and duration of the task, possible failures, and recovery strategies. Document any contextual factors in the HRI study that are different from the envisioned use cases.

3. **Design**: Assemble a diverse design team and/or advisory board that includes researchers from related disciplines, domain experts, primary users, and other stakeholders. Conduct an iterative and collaborative design process to build a shared vocabulary in this diverse team. Have a tangible realisation of the design to communicate both the capabilities and limitations of the HRI system.

4. **Evaluation**: Broaden the scope of evaluation to incorporate approaches from other disciplines supported by your diverse team. When using quantitative methods to evaluate a factory, support your result by considering multiple metrics. Employ qualitative and mixed methods to gain understanding of subjective and contextualised experience in an interaction.

5. **Stakeholders**: Clarify the roles of the humans in an HRI study and in the application of an HRI system. Design and evaluate the

HRI system according to the user context. Establish a mutual communication and adaption between the development end and the reception end of the HRI system.

6. **Data**: Statistical power is important, but does not trump all, in HRI studies. Choose suitable data-processing and analysis approaches according to the experiment and evaluation design. Incorporate the study contexts when interpreting observations.

7. **Ethics**: Evaluate whether the research questions and goals satisfy ethical considerations. Apply a social and cultural lens to develop beneficial and acceptable HRI systems. Document possible biases that can arise from design, evaluation, experimental setups, or data analysis. Address potential limits that can be caused by the scope of the design team or the demographics of the participants. Account for possible risks and misuses of the HRI systems by establishing transparent and bidirectional communication and control channels between designers and users.

Bibliography

Hamid Abbasi-Bafghi, Hamid R Fallah-Yakhdani, Onno G Meijer, Henrica CW de Vet, Sjoerd M Bruijn, Li-Yong Yang, Dirk L Knol, Barend J Van Royen, and Jaap H van Dieën. The effects of knee arthroplasty on walking speed: A meta-analysis. *BMC Musculoskeletal Disorders*, 13(1):66, 2012. doi: 10.1186/1471-2474-13-66.

Moaed A Abd, Iker Gonzalez, Mehrdad Nojoumian, and Erik D Engeberg. Trust, satisfaction and frustration measurements during human-robot interaction. In *30th Florida Conference on Recent Advances in Robotics, May*, pages 11–12, 2017.

Elahe Abdi, Mohamed Bouri, Jeremy Olivier, and Hannes Bleuler. Foot-controlled endoscope positioner for laparoscopy: Development of the master and slave interfaces. In *2016 4th International Conference on Robotics and Mechatronics (ICROM)*, pages 111–115. IEEE, 2016.

Martha Abshire, Victor D Dinglas, Maan Isabella A Cajita, Michelle N Eakin, Dale M Needham, and Cheryl Dennison Himmelfarb. Participant retention practices in longitudinal clinical research studies with high retention rates. *BMC Medical Research Methodology*, 17(1):1–10, 2017.

Josh Achiam, Steven Adler, Sandhini Agarwal, Lama Ahmad, Ilge Akkaya, Florencia Leoni Aleman, Diogo Almeida, Janko Altenschmidt, Sam Altman, Shyamal Anadkat, et al. GPT-4 technical report. *arXiv preprint arXiv:2303.08774*, 2023.

Henny Admoni and Brian Scassellati. Social eye gaze in human-robot interaction: A review. *Journal of Human-Robot Interaction*, 6(1):25, 2017. doi: 10.5898/JHRI.6.1.Admoni.

Arvin Agah. Human interactions with intelligent systems: research taxonomy. *Computers & Electrical Engineering*, 27(1):71–107, 2000. doi: 10.1016/S0045-7906(00)00009-4.

Nick Ah Sen, Pamela Carreno-Medrano, and Dana Kulić. Human-aware subgoal generation in crowded indoor environments. In *International Conference on Social Robotics*, pages 50–60. Springer, 2022.

Muneeb Imtiaz Ahmad, Omar Mubin, and Joanne Orlando. Adaptive social robot for sustaining social engagement during long-term children-robot interaction. *International Journal of Human-Computer Interaction*, 33(12):943–962, 2017. doi: 10.1080/10447318.2017.1300750.

Neziha Akalin, Annica Kristoffersson, and Amy Loutfi. The influence of feedback type in robot-assisted training. *Multimodal Technologies and Interaction*, 3(4):67, 2019.

Baris Akgun and Kaushik Subramanian. Robot learning from demonstration: Kinesthetic teaching vs. teleoperation, 2011. URL https://api.semanticscholar.org/CorpusID:45150299.

Minoo Alemi, Ali Meghdari, and Nafiseh Sadat Haeri. Young EFL learners' attitude towards RALL: An observational study focusing on motivation, anxiety, and interaction. In *International Conference on Social Robotics*, pages 252–261. Springer, 2017.

R. McN. Alexander. The gaits of bipedal and quadrupedal animals. *The International Journal of Robotics Research*, 3(2):49–59, 1984. doi: 10.1177/027836498400300205.

Aimee Allen, Tom Drummond, and Dana Kulić. Effects of consequential sounds on human-perception of robots (supplementary website). https://aimeeallen.github.io/perception-of-robot-consequential-sounds/, 2024a. Accessed: 2024-10-24.

Aimee Allen, Tom Drummond, and Dana Kulić. Robots have been seen and not heard: Effects of consequential sounds on human-perception of robots. *arXiv preprint arXiv:2406.02938*, 2024b.

Mike Allen. *The SAGE encyclopedia of communication research methods*. Sage Publications, 2017.

Naomi Altman and Martin Krzywinski. Points of significance: interpreting *P* values, *Nature Methods*, 14:213–214, 2017.

Patrícia Alves-Oliveira, Patrícia Arriaga, Ana Paiva, and Guy Hoffman. Children as robot designers. In *Proceedings of the 2021 ACM/IEEE International Conference on Human-Robot Interaction*, pages 399–408, 2021.

Salvatore M Anzalone, Sofiane Boucenna, Serena Ivaldi, and Mohamed Chetouani. Evaluating the engagement with social robots. *International Journal of Social Robotics*, 7(4):465–478, 2015.

Tamio Arai, Ryu Kato, and Marina Fujita. Assessment of operator stress induced by robot collaboration in assembly. *CIRP Annals*, 59(1):5–8, 2010.

Stephanie Arévalo Arboleda, Max Pascher, Younes Lakhnati, and Jens Gerken. Understanding human-robot collaboration for people with mobility impairments at the workplace, a thematic analysis. In *2020 29th IEEE International Conference on Robot and Human Interactive Communication (RO-MAN)*, pages 561–566. IEEE, 2020.

Stephanie Arevalo Arboleda, Max Pascher, Annalies Baumeister, Barbara Klein, and Jens Gerken. Reflecting upon participatory design in human-robot collaboration for people with motor disabilities: Challenges and lessons learned from three multiyear projects. In *The 14th PErvasive Technologies Related to Assistive Environments Conference*, pages 147–155, 2021.

Brenna D. Argall, Sonia Chernova, Manuela Veloso, and Brett Browning. A survey of robot learning from demonstration. *Robotics and Autonomous Systems*, 57(5):469–483, 2009. doi: 10.1016/j.robot.2008.10.024.

RA Armstrong and Anthony Hilton. The use of analysis of variance (ANOVA) in applied microbiology. *Microbiologist*, 5(4):18–21, 2004.

Minja Axelsson, Raquel Oliveira, Mattia Racca, and Ville Kyrki. Social robot co-design canvases: A participatory design framework. *ACM Transactions on Human-Robot Interaction (THRI)*, 11(1):1–39, 2021.

Minja Axelsson, Micol Spitale, and Hatice Gunes. Robots as mental well-being coaches: Design and ethical recommendations. *ACM Transactions on Human-Robot Interaction*, 13(2):1–55, 2024.

Franziska Babel, Johannes Kraus, Philipp Hock, Hannah Asenbauer, and Martin Baumann. Investigating the validity of online robot evaluations: Comparison of findings from an one-sample online and laboratory study. In *Companion of the 2021 ACM/IEEE International Conference on Human-Robot Interaction*, pages 116–120, 2021.

Kal Backman, Dana Kulić, and Hoam Chung. Reinforcement learning for shared autonomy drone landings. *Autonomous Robots*, 47(8):1419–1438, 2023.

Shelly Bagchi, Patrick Holthaus, Gloria Beraldo, Emmanuel Senft, Daniel Hernandez Garcia, Zhao Han, Suresh Kumaar Jayaraman, Alessandra Rossi, Connor Esterwood, Antonio Andriella, et al. Towards improved replicability of human studies in human-robot interaction: Recommendations for formalized reporting. In *Companion of the 2023 ACM/IEEE International Conference on Human-Robot Interaction*, pages 629–633, 2023.

Lisanne Bainbridge. Ironies of automation. In *Analysis, design and evaluation of man–machine systems*, pages 129–135. Elsevier, 1983.

Jimmy Baraglia, Maya Cakmak, Yukie Nagai, Rajesh PN Rao, and Minoru Asada. Efficient human-robot collaboration: When should a robot take initiative? *The International Journal of Robotics Research*, 36(5–7):563–579, 2017. doi: 10.1177/0278364916688253.

Lisa Feldman Barrett, Ralph Adolphs, Stacy Marsella, Aleix M Martinez, and Seth D Pollak. Emotional expressions reconsidered: Challenges to inferring emotion from human facial movements. *Psychological Science in the Public Interest*, 20(1):1–68, 2019.

C. Bartneck, T. Belpaeme, F. Eyssel, T. Kanda, M. Keijsers, and S. Sabanovic. *Human-Robot Interaction – An Introduction.* Cambridge: Cambridge University Press, 2020a.

Christoph Bartneck, Dana Kulić, Elizabeth Croft, and Susana Zoghbi. Measurement instruments for the anthropomorphism, animacy, likeability, perceived intelligence, and perceived safety of robots. *International Journal of Social Robotics*, 1(1):71–81, 2009.

Christoph Bartneck, Kumar Yogeeswaran, Qi Min Ser, Graeme Woodward, Robert Sparrow, Siheng Wang, and Friederike Eyssel. Robots and racism. In *Proceedings of the 2018 ACM/IEEE International Conference on Human-Robot Interaction*, pages 196–204, 2018.

Christoph Bartneck, Tony Belpaeme, Friederike Eyssel, Takayuki Kanda, Merel Keijsers, and Selma Šabanović. *Human-Robot Interaction: An Introduction.* Cambridge University Press, 2020b.

Saša Baškarada. Qualitative case study guidelines. *The Qualitative Report*, 19(40):1–25, 2014.

Paul Baxter, James Kennedy, Emmanuel Senft, Severin Lemaignan, and Tony Belpaeme. From characterising three years of HRI to methodology and reporting recommendations. In *2016 11th ACM/IEEE International Conference on Human-Robot Interaction (HRI)*, pages 391–398. IEEE, 2016.

Tomislav Baček, Mingrui Sun, Hengchang Liu, Zhongxiang Chen, Chris Manzie, Etienne Burdet, Dana Kulić, Denny Oetomo, and Ying Tan. A biomechanics and energetics dataset of neurotypical adults walking with and without kinematic constraints. *Scientific Data*, 11(1):646, 2024. doi: 10.1038/s41597-024-03444-4.

Jenay M Beer, Arthur D Fisk, and Wendy A Rogers. Toward a framework for levels of robot autonomy in human-robot interaction. *Journal of Human-Robot Interaction*, 3(2):74, 2014. doi: 10.5898/JHRI.3.2.Beer.

Kathleen Belhassein, Guilhem Buisan, Aurélie Clodic, and Rachid Alami. Towards methodological principles for user studies in human-robot interaction. In *Test Methods and Metrics for Effective HRI in Collaborative Human-Robot Teams Workshop, ACM/IEEE International Conference on Human-Robot Interaction*, 2019.

Rachel K. E. Bellamy, Kuntal Dey, Michael Hind, Samuel C. Hoffman, Stephanie Houde, Kalapriya Kannan, Pranay Lohia, Jacquelyn Martino, Sameep Mehta, Aleksandra Mojsilovic, Seema Nagar, Karthikeyan Natesan Ramamurthy, John Richards, Diptikalyan Saha, Prasanna Sattigeri, Moninder Singh, Kush R. Varshney, and Yunfeng Zhang. AI Fairness 360: An extensible toolkit for detecting, understanding, and mitigating unwanted algorithmic bias, October 2018. https://arxiv.org/abs/ 1810.01943.

Elaine M Beller, Val Gebski, and Anthony C Keech. Randomisation in clinical trials. *Medical Journal of Australia*, 177(10):565–567, 2002.

Marcus Bendtsen. Avoiding under- and over recruitment in behavioral intervention trials using Bayesian sequential designs: tutorial. *Journal of Medical Internet Research*, 24(12):e40730, 2022.

Ruha Benjamin. Which humans? Innovation, equity, and imagination in human-centered design (keynote). https://www.youtube.com/watch?v= kDcz44ifdQw, May 2021.

Nicholas Berente, Uri Gal, and Sean Hansen. Ethical implications of social stratification in information systems research. *Information Systems Journal*, 21(4):357–382, 2011.

Nicole Bergen and Ronald Labonté. "Everything is perfect, and we have no problems": detecting and limiting social desirability bias in qualitative research. *Qualitative Health Research*, 30(5):783–792, 2020.

H Russell Bernard. *Research methods in anthropology: Qualitative and quantitative approaches*. Rowman & Littlefield, 2017.

Henning Best and Christof Wolf. *The SAGE handbook of regression analysis and causal inference*. Sage, 2014.

Alberto Betella and Paul FMJ Verschure. The affective slider: a digital self-assessment scale for the measurement of human emotions. *PLoS One*, 11(2), 2016.

Cindy L Bethel. *Robots without faces: non-verbal social human-robot interaction*. University of South Florida, 2009.

Cindy L Bethel and Robin R Murphy. Review of human studies methods in HRI and recommendations. *International Journal of Social Robotics*, 2(4): 347–359, 2010.

Cindy L Bethel, Kristen Salomon, Robin R Murphy, and Jennifer L Burke. Survey of psychophysiology measurements applied to human-robot interaction. In *RO-MAN 2007-The 16th IEEE International Symposium on Robot and Human Interactive Communication*, pages 732–737. IEEE, 2007.

Cindy L Bethel, Jessie E Cossitt, Zachary Henkel, and Kenna Baugus. Qualitative interview techniques for human-robot interactions. In *Human-Robot Interaction*, pages 145–174. Springer, 2020a.

Cindy L Bethel, Zachary Henkel, and Kenna Baugus. Conducting studies in human-robot interaction. In *Human-Robot Interaction*, pages 91–124. Springer, 2020b.

Elin A. Björling, Honson Ling, Simran Bhatia, and Kimberly Dziubinski. The experience and effect of adolescent to robot stress disclosure: A mixed-methods exploration. In Alan R. Wagner, David Feil-Seifer, Kerstin S. Haring, Silvia Rossi, Thomas Williams, Hongsheng He, and Shuzhi Sam Ge, editors, *Social Robotics*, pages 604–615. Springer International Publishing, Cham, 2020.

María J Blanca, Rafael Alarcón, Jaume Arnau, Roser Bono, and Rebecca Bendayan. Non-normal data: is ANOVA still a valid option? *Psicothema*, 29(4):552–557, 2017.

Vincent Blok. *Putting Responsible Research and Innovation into Practice: A Multi-Stakeholder Approach.* Springer, 2022.

Lasse Blond. Studying robots outside the lab: HRI as ethnography. *Paladyn, Journal of Behavioral Robotics*, 10(1):117–127, 2019.

Júlia Pareto Boada, Begoña Román Maestre, and Carme Torras Genís. The ethical issues of social assistive robotics: A critical literature review. *Technology in Society*, 67:101726, 2021.

Indu P Bodala, Bing Cai Kok, Weicong Sng, and Harold Soh. Modeling the interplay of trust and attention in HRI: An autonomous vehicle study. In *Companion of the 2020 ACM/IEEE International Conference on Human-Robot Interaction*, pages 145–147, 2020.

Margaret Boden, Joanna Bryson, Darwin Caldwell, Kerstin Dautenhahn, Lilian Edwards, Sarah Kember, Paul Newman, Vivienne Parry, Geoff Pegman, Tom Rodden, et al. Principles of robotics: regulating robots in the real world. *Connection Science*, 29(2):124–129, 2017.

Susanne Bødker and Morten Kyng. Participatory design that matters—facing the big issues. *ACM Transactions on Computer-Human Interaction (TOCHI)*, 25(1):1–31, 2018.

Susanne Bødker, Christian Dindler, and Ole Sejer Iversen. Tying knots: Participatory infrastructuring at work. *Computer Supported Cooperative Work (CSCW)*, 26(1-2):245–273, 2017.

Dan Bohus, Chit W Saw, and Eric Horvitz. Directions robot: in-the-wild experiences and lessons learned. In *Proceedings of the 2014 International Conference on Autonomous Agents and Multi-Agent Systems*, pages 637–644. Citeseer, 2014.

Fabio Bonsignorio and Angel P Del Pobil. Toward replicable and measurable robotics research [from the guest editors]. *IEEE Robotics & Automation Magazine*, 22(3):32–35, 2015.

Margaret M Bradley and Peter J Lang. Measuring emotion: the self-assessment manikin and the semantic differential. *Journal of Behavior Therapy and Experimental Psychiatry*, 25(1):49–59, 1994.

Eva Brandt. Designing exploratory design games: a framework for participation in participatory design? In *Proceedings of the ninth conference on Participatory design: Expanding boundaries in design-Volume 1*, pages 57–66, 2006.

Virginia Braun and Victoria Clarke. Using thematic analysis in psychology. *Qualitative research in psychology*, 3(2):77–101, 2006.

C. Breazeal, C.D. Kidd, A.L. Thomaz, G. Hoffman, and M. Berlin. Effects of nonverbal communication on efficiency and robustness in human-robot teamwork. In *2005 IEEE/RSJ International Conference on Intelligent Robots and Systems*, pages 708–713, 2005. doi: 10.1109/IROS.2005. 1545011.

Marc Brysbaert. How many participants do we have to include in properly powered experiments? A tutorial of power analysis with reference tables. *Journal of Cognition*, 2(1), 2019.

John M Bryson. What to do when stakeholders matter: stakeholder identification and analysis techniques. *Public Management Review*, 6(1):21–53, 2004.

Christina Bröhl, Jochen Nelles, Christopher Brandl, Alexander Mertens, and Verena Nitsch. Human-robot collaboration acceptance model: Development and comparison for germany, japan, china and the usa. *International Journal of Social Robotics*, 11(5):709–726, 2019. doi: 10.1007/s12369-019-00593-0.

Charlotte Burck. Comparing qualitative research methodologies for systemic research: the use of grounded theory, discourse analysis and narrative analysis. *Journal of Family Therapy*, 27(3):237–262, 2005.

Dafna Burema. A critical analysis of the representations of older adults in the field of human–robot interaction. *AI & SOCIETY*, pages 1–11, 2021.

Loraine Busetto, Wolfgang Wick, and Christopg Gumbinger. How to use and assess qualitative research methods. *Neurological Research and Practice*, 2, 2020.

Martina Čaić, Joao Avelino, Dominik Mahr, Gaby Odekerken-Schröder, and Alexandre Bernardino. Robotic versus human coaches for active aging: An automated social presence perspective. *International Journal of Social Robotics*, 12(4):867–882, 2020.

Praminda Caleb-Solly, Sanja Dogramadzi, David Ellender, Tina Fear, and Herjan van den Heuvel. A mixed-method approach to evoke creative and holistic thinking about robots in a home environment. In *Proceedings of the 2014 ACM/IEEE International Conference on Human-Robot Interaction*, pages 374–381, 2014.

Rudolf N Cardinal and Michael RF Aitken. *ANOVA for the behavioral sciences researcher*. Psychology Press, 2013.

Colleen M Carpinella, Alisa B Wyman, Michael A Perez, and Steven J Stroessner. The robotic social attributes scale (RoSAS) development and validation. In *Proceedings of the 2017 ACM/IEEE International Conference on Human-Robot Interaction*, pages 254–262, 2017.

Pamela Carreno-Medrano, Leimin Tian, Aimee Allen, Shanti Sumartojo, Michael Mintrom, Enrique Coronado, Gentiane Venture, Elizabeth Croft, and Dana Kulić. Aligning robot's behaviours and users' perceptions through participatory prototyping. *arXiv preprint arXiv:2101.03660*, 2021.

Pamela Carreno-Medrano, Leimin Tian, Aimee Allen, Shanti Sumartojo, Enrique Coronado, Gentiane Venture, Elizabeth Croft, and Dana Kulić. A participatory design methodology for building shared understanding between users and designers in HRI. *arXiv preprint arXiv:2101.03660*, 2024.

Felix Carros, Johanna Meurer, Diana Löffler, David Unbehaun, Sarah Matthies, Inga Koch, Rainer Wieching, Dave Randall, Marc Hassenzahl, and Volker Wulf. Exploring human-robot interaction with the elderly: Results from a ten-week case study in a care home. In *Proceedings of the 2020 CHI Conference on Human Factors in Computing Systems*, CHI '20, pages 1–12. Association for Computing Machinery, 2020.

Dylan Cawthorne. Robot ethics: Ethical design considerations. In *Foundations of Robotics*, pages 473–491. Springer, 2022.

Jessy Ceha, Nalin Chhibber, Joslin Goh, Corina McDonald, Pierre-Yves Oudeyer, Dana Kulić, and Edith Law. Expression of curiosity in social robots: Design, perception, and effects on behaviour. In *Proceedings of the 2019 CHI Conference on Human Factors in Computing Systems*, pages 1–12, 2019.

Nathalia Cespedes, Marcela Munera, Catalina Gomez, and Carlos A Cifuentes. Social human-robot interaction for gait rehabilitation. *IEEE Transactions on Neural Systems and Rehabilitation Engineering*, 28(6):1299–1307, 2020.

Wesley P Chan, Chris AC Parker, HF Machiel Van der Loos, and Elizabeth A Croft. A human-inspired object handover controller. *The International Journal of Robotics Research*, 32(8):971–983, 2013. doi: 10.1177/0278364913488806.

Wesley P. Chan, Geoffrey Hanks, Maram Sakr, Tiger Zuo, H. F. Machiel Van der Loos, and Elizabeth A. Croft. An augmented reality human-robot physical collaboration interface design for shared, large-scale, labour-intensive manufacturing tasks. *IROS*, 2020.

Jesse Chandler, Pam Mueller, and Gabriele Paolacci. Nonnaïveté among Amazon Mechanical Turk workers: Consequences and solutions for behavioral researchers. *Behavior research methods*, 46:112–130, 2014.

Jesse Chandler, Gabriele Paolacci, Eyal Peer, Pam Mueller, and Kate A Ratliff. Using nonnaive participants can reduce effect sizes. *Psychological science*, 26(7):1131–1139, 2015.

Wan-Ling Chang and Selma Šabanović. Interaction expands function: Social shaping of the therapeutic robot PARO in a nursing home. In *2015 10th ACM/IEEE International Conference on Human-Robot Interaction (HRI)*, pages 343–350. IEEE, 2015.

Wan-Ling Chang, Selma Šabanović, and Lesa Huber. Observational study of naturalistic interactions with the socially assistive robot PARO in a nursing home. In *The 23rd IEEE International Symposium on Robot and Human Interactive Communication*, pages 294–299. IEEE, 2014.

Kathy Charmaz. *Constructing grounded theory: A practical guide through qualitative analysis*. Sage, 2006.

Samprit Chatterjee and Jeffrey S. Simonoff. *Handbook of Regression Analysis*. Wiley, 2012.

Ronald J Chenail. Interviewing the investigator: Strategies for addressing instrumentation and researcher bias concerns in qualitative research. *Qualitative Report*, 16(1):255–262, 2011.

Ji Young Cho and Eun-Hee Lee. Reducing confusion about grounded theory and qualitative content analysis: Similarities and differences. *Qualitative Report*, 19(32), 2014.

Youngjoon Choi, Miju Choi, Munhyang Oh, and Seongseop Kim. Service robots in hotels: understanding the service quality perceptions of human-robot interaction. *Journal of Hospitality Marketing & Management*, 29(6):613–635, 2020.

Larry B. Christensen. *Experimental methodology.* Pearson/Allyn & Bacon, Boston, 10th ed. 2007.

Michael Jae-Yoon Chung. Awesome HRI datasets and human simulators. https://github.com/mjyc/awesome-hri-datasets, 2021. Accessed: 2024-09-04.

Tom Clark. On 'being researched': Why do people engage with qualitative research? *Qualitative Research*, 10(4):399–419, 2010.

Responsible AI Collaborative. AI incident database. https://incidentdatabase.ai/, 2024. Accessed: 2024-08-09.

Emily C. Collins. Drawing parallels in human-other interactions: a trans-disciplinary approach to developing human-robot interaction methodologies. *Philosophical Transactions of the Royal Society of London. Series B, Biological Sciences*, 374(1771):20180433, 2019. doi: 10.1098/rstb.2018.0433.

Enrique Coronado, Dominique Deuff, Pamela Carreno-Medrano, Leimin Tian, Dana Kulić, Shanti Sumartojo, Fulvio Mastrogiovanni, and Gentiane Venture. Towards a modular and distributed end-user development framework for human-robot interaction. *IEEE Access*, 9:12675–12692, 2021.

Filipa Correia, Carla Guerra, Samuel Mascarenhas, Francisco S Melo, and Ana Paiva. Exploring the impact of fault justification in human-robot trust. In *Proceedings of the 17th International Conference on Autonomous Agents and MultiAgent Systems*, pages 507–513, 2018.

António Pedro Costa, Francislê Neri de Sousa, António Moreira, and Dayse Neri de Souza. Research through design: qualitative analysis to evaluate the usability. In *Computer supported qualitative research*, pages 1–12. Springer, 2017.

Sasha Costanza-Chock. *Design justice: Community-led practices to build the worlds we need.* The MIT Press, 2020.

Elisabeth Coyne, Eileen Grafton, and Alayne Reid. Strategies to successfully recruit and engage clinical nurses as participants in qualitative clinical research. *Contemporary Nurse*, 52(6):669–676, 2016.

Denis Cristol and Dorothée Cavignaux-Bros. Participatory design and co-design-The case of a MOOC on public innovation. *Learner and User Experience Research*, 2020.

Geoff Cumming. The new statistics: Why and how. *Psychological science*, 25 (1):7–29, 2014.

Praveen Damacharla, Ahmad Y Javaid, Jennie J Gallimore, and Vijay K Devabhaktuni. Common metrics to benchmark human-machine teams (HMT): a review. *IEEE Access*, 6:38637–38655, 2018.

Kerstin Dautenhahn. Methodology & themes of human-robot interaction: A growing research field. *International Journal of Advanced Robotic Systems*, 4(1):15, 2007.

Kerstin Dautenhahn. Some brief thoughts on the past and future of human-robot interaction. *ACM Transactions on Human-Robot Interaction (THRI)*, 7(1):1–3, 2018.

Kerstin Dautenhahn and Joe Saunders. *New Frontiers in Human Robot Interaction*, volume 2, John Benjamins Publishing, 2011.

Laurens De Gauquier, Malaika Brengman, Kim Willems, Hoang-Long Cao, and Bram Vanderborght. In or out? a field observational study on the placement of entertaining robots in retailing. *International Journal of Retail & Distribution Management*, 49(7):846–874, 2021.

Maartje MA de Graaf. An ethical evaluation of human–robot relationships. *International Journal of Social Robotics*, 8(4):589–598, 2016.

Maartje MA de Graaf, Somaya Ben Allouch, and Jan AGM van Dijk. Long-term evaluation of a social robot in real homes. *Interaction Studies*, 17(3):462–491, 2016.

Scott L. Delp, Frank C. Anderson, Allison S. Arnold, Peter Loan, Ayman Habib, Chand T. John, Eran Guendelman, and Darryl G. Thelen. OpenSim: Open-source software to create and analyze dynamic simulations of movement. *IEEE Transactions on Biomedical Engineering*, 54(11):1940–1950, 2007. doi: 10.1109/TBME.2007.901024.

DFAT. Ethical research and evaluation guidance note. https://www.dfat.gov.au/sites/default/files/ethical-research-evaluation-guidance-note.pdf, 2021. Accessed: 2024-08-09.

Amir Dini, Cornelia Murko, Saeed Yahyanejad, Ursula Augsdörfer, Michael Hofbaur, Lucas Paletta. Measurement and prediction of situation awareness in human-robot interaction based on a framework of probabilistic attention. In *2017 IEEE/RSJ International Conference on Intelligent Robots and Systems (IROS)*, pages 4354–4361. IEEE, 2017.

Carl DiSalvo, Marti Louw, David Holstius, Illah Nourbakhsh, and Ayca Akin. Toward a public rhetoric through participatory design: Critical engagements and creative expression in the neighborhood networks project. *Design Issues*, 28(3):48–61, 2012.

Alan Dix. *Statistics for HCI: Making Sense of Quantitative Data*. Morgan & Claypool, 2020.

Gordon S Doig and Fiona Simpson. Randomization and allocation concealment: a practical guide for researchers. *Journal of Critical Care*, 20(2):187–191, 2005.

Birsen Donmez, Patricia E Pina, and Mary L Cummings. Evaluation criteria for human-automation performance metrics. In *Performance Evaluation and Benchmarking of Intelligent Systems*, pages 21–40. Springer, 2009.

Anca D Dragan, Shira Bauman, Jodi Forlizzi, and Siddhartha S Srinivasa. Effects of robot motion on human-robot collaboration. In *2015 10th ACM/IEEE International Conference on Human-Robot Interaction (HRI)*, pages 51–58. IEEE, 2015.

David L Driscoll, Afua Appiah-Yeboah, Philip Salib, and Douglas J Rupert. Merging qualitative and quantitative data in mixed methods research: How to and why not. *Ecological and Environmental Anthropology*, 2007.

Ezequiel Duque, Guilherme Fonseca, Heitor Vieira, Gustavo Gontijo, and Lucila Ishitani. A systematic literature review on user centered design and participatory design with older people. In *Proceedings of the 18th Brazilian Symposium on Human Factors in Computing Systems*, pages 1–11. Association for Computing Machinery, 2019.

Bosede I. Edwards, Idris O. Muniru, Nosiba Khougali, Adrian D. Cheok, and Rui Prada. A physically embodied robot teacher (PERT) as a facilitator for peer learning. In *2018 IEEE Frontiers in Education Conference (FIE)*, pages 1–9. IEEE, 2018. doi: 10.1109/FIE.2018.8658445.

Carsten Eickhoff and Arjen P de Vries. Increasing cheat robustness of crowdsourcing tasks. *Information Retrieval*, 16(2):121–137, 2013.

Michael Eliasziw, S Lorraine Young, M Gail Woodbury, and Karen Fryday-Field. Statistical methodology for the concurrent assessment of interrater and intrarater reliability: using goniometric measurements as an example. *Physical Therapy*, 74(8):777–788, 1994.

Satu Elo and Helvi Kyngäs. The qualitative content analysis process. *Journal of Advanced Nursing*, 62(1):107–115, 2008.

Mica R Endsley, Stephen J Selcon, Thomas D Hardiman, and Darryl G Croft. A comparative analysis of SAGAT and SART for evaluations of situation awareness. In *Proceedings of the Human Factors and Ergonomics Society Annual Meeting*, volume 42, pages 82–86. SAGE Publications, 1998.

David M Erceg-Hurn and Vikki M Mirosevich. Modern robust statistical methods: an easy way to maximize the accuracy and power of your research. *American Psychologist*, 63(7):591, 2008.

Pablo Estefo, Jocelyn Simmonds, Romain Robbes, and Johan Fabry. The robot operating system: Package reuse and community dynamics. *Journal of Systems and Software*, 151:226–242, 2019.

Connor Esterwood and Lionel Peter Robert Jr. The warehouse robot interaction Sim: An open-source HRI research platform. In *Companion of the 2023 ACM/IEEE International Conference on Human-Robot Interaction*, pages 268–271, 2023.

Ilker Etikan, Sulaiman Abubakar Musa, and Rukayya Sunusi Alkassim. Comparison of convenience sampling and purposive sampling. *American Journal of Theoretical and Applied Statistics*, 5(1):1–4, 2016.

Alexander Etz and Joachim Vandekerckhove. Introduction to Bayesian inference for psychology. *Psychonomic Bulletin & Review*, 25(1):5–34, 2018.

Alexander Etz, Quentin F Gronau, Fabian Dablander, Peter A Edelsbrunner, and Beth Baribault. How to become a Bayesian in eight easy steps: An annotated reading list. *Psychonomic Bulletin & Review*, 25(1):219–234, 2018.

Hugo Fastl and Eberhard Zwicker. Masking. In *Psychoacoustics: Facts and Models*, pages 61–110. Springer, 2007. doi: 10.4324/ 9781843143598-58.

Franz Faul, Edgar Erdfelder, Albert-Georg Lang, and Axel Buchner. G* Power 3: A flexible statistical power analysis program for the social, behavioral, and biomedical sciences. *Behavior Research Methods*, 39(2):175–191, 2007.

Joe R Feagin, Anthony M Orum, and Gideon Sjoberg. *A Case for the Case Study*. UNC Press Books, 1991.

David Feil-Seifer and Maja J Mataric. Human robot interaction. *Encyclopedia of Complexity and Systems Science*, 80:4643–4659, 2009.

Betty J Feir-Walsh and Larry E Toothaker. An empirical comparison of the ANOVA F-test, normal scores test and Kruskal-Wallis test under violation of assumptions. *Educational and Psychological Measurement*, 34(4):789–799, 1974.

Ylva Fernaeus, Maria Håkansson, Mattias Jacobsson, and Sara Ljungblad. How do you play with a robotic toy animal? A long-term study of Pleo. In *Proceedings of the 9th international Conference on interaction Design and Children*, pages 39–48. Association for Computing Machinery, 2010.

Linda Finlay. Engaging phenomenological analysis. *Qualitative Research in Psychology*, 11(2):121–141, 2014.

Kerstin Fischer. Effect confirmed, patient dead: A commentary on Hoffman & Zhao's Primer for Conducting Experiments in HRI. *ACM Transactions on Human-Robot Interaction (THRI)*, 10(1):1–4, 2021.

Kerstin Fischer, Katrin Lohan, Joe Saunders, Chrystopher Nehaniv, Britta Wrede, and Katharina Rohlfing. The impact of the contingency of robot feedback on HRI. In *2013 International Conference on Collaboration Technologies and Systems (CTS)*, pages 210–217. IEEE, 2013.

Uwe Flick. Mapping the field. In Uwe Flick, editor, *The SAGE Handbook of Qualitative Data Analysis*, pages 4–18. Sage, 2013.

Jodi Forlizzi, Carl DiSalvo, and Francine Gemperle. Assistive robotics and an ecology of elders living independently in their homes. *Human–Computer Interaction*, 19(1–2):25–59, 2004.

Thomas Franke, Christiane Attig, and Daniel Wessel. A personal resource for technology interaction: development and validation of the affinity for technology interaction (ATI) scale. *International Journal of Human–Computer Interaction*, 35(6):456–467, 2019.

Christopher Frauenberger, Marjo Rauhala, and Geraldine Fitzpatrick. In-action ethics. *Interacting with Computers*, 29(2):220–236, 2017.

Marlena R Fraune, Satoru Kawakami, Selma Šabanović, P Ravindra S De Silva, and Michio Okada. Three's company, or a crowd? The effects of robot number and behavior on HRI in Japan and the USA. In *Robotics: Science and Systems*. MIT Press Journals, 2015.

Morten Roed Frederiksen and Kasper Stoey. Augmenting the audio-based expression modality of a non-affective robot. In *2019 8th International Conference on Affective Computing and Intelligent Interaction (ACII)*, pages 144–149, 2019. doi: 10.1109/ACII.2019.8925510.

Susanne Frennert, Håkan Eftring, and Britt Östlund. Case report: Implications of doing research on socially assistive robots in real homes. *International Journal of Social Robotics*, 9(3):401–415, 2017.

Rudolf Freund, Donna Mohr, and William Wilson. *Statistical Methods*. Elsevier, 2010.

Emma Frid, Roberto Bresin, and Simon Alexanderson. Perception of mechanical sounds inherent to expressive gestures of a NAO robot-implications for movement sonification of humanoids. In *Sound and Music Computing*, pages 43–51, Zenodo Cyprus, 2018.

Batya Friedman, David G Hendry, Alan Borning, et al. A survey of value sensitive design methods. *Foundations and Trends® in Human–Computer Interaction*, 11(2):63–125, 2017.

Craig L Fry, Wayne Hall, Alison Ritter, and Rebecca Jenkinson. The ethics of paying drug users who participate in research: A review and practical recommendations. *Journal of Empirical Research on Human Research Ethics*, 1(4):21–35, 2006.

Scott Fujimoto, Herke Hoof, and David Meger. Addressing function approximation error in actor-critic methods. In *International Conference on Machine Learning*, pages 1587–1596. PMLR, 2018.

David C Funder and Daniel J Ozer. Evaluating effect size in psychological research: Sense and nonsense. *Advances in Methods and Practices in Psychological Science*, 2(2):156–168, 2019.

Patricia I Fusch, Gene E Fusch, and Lawrence R Ness. How to conduct a mini-ethnographic case study: A guide for novice researchers. *The Qualitative Report*, 22(3):923–942, 2017.

Martin J Gardner and Douglas G Altman. Confidence intervals rather than p values: estimation rather than hypothesis testing. *British Medical Journal (Clinicl Research Ed)*, 292(6522):746–750, 1986.

Ilaria Gaudiello, Elisabetta Zibetti, Sébastien Lefort, Mohamed Chetouani, and Serena Ivaldi. Trust as indicator of robot functional and social acceptance. An experimental study on user conformation to ICUB answers. *Computers in Human Behavior*, 61:633–655, 2016.

Andrew Gelman. Ethics in statistical practice and communication: Five recommendations. *Significance*, 15(5):40–43, 2018.

Christopher H George, S Clare Stanford, Steve Alexander, Giuseppe Cirino, James R Docherty, Mark A Giembycz, Daniel Hoyer, Paul A Insel, Angelo A Izzo, Yong Ji, et al. Updating the guidelines for data transparency in the British Journal of Pharmacology – data sharing and the use of scatter plots instead of bar charts. *British Journal of Pharmacology*, 174(17):2801, 2017.

Darren Gergle and Desney S Tan. Experimental research in HCI. In *Ways of Knowing in HCI*, pages 191–227. Springer, 2014.

Aimi S Ghazali, Jaap Ham, Emilia I Barakova, and Panos Markopoulos. Effects of robot facial characteristics and gender in persuasive human-robot interaction. *Frontiers in Robotics and AI*, 5:73, 2018.

Michael J. Gielniak and Andrea L. Thomaz. Enhancing interaction through exaggerated motion synthesis. In *2012 7th ACM/IEEE International Conference on Human-Robot Interaction (HRI)*, pages 375–382. IEEE, 2012. doi: 10.1145/2157689.2157813.

Curtis L. Gittens. Remote HRI: A methodology for maintaining COVID-19 physical distancing and human interaction requirements in HRI studies. *Information Systems Frontiers*, 26:1–16, 2021.

Michael Goodrich and Alan Schultz. Human-robot interaction: A survey. *Foundations and Trends in Human-Computer Interaction*, 1:203–275, 2007. . doi: 10.1561/1100000005.

A. Green and K.S. Eklundh. Designing for learnability in human-robot communication. *IEEE Transactions on Industrial Electronics*, 50(4):644–650, 2003. doi: 10.1109/TIE.2003.814763.

Robert J Grissom. Heterogeneity of variance in clinical data. *Journal of Consulting and Clinical Psychology*, 68(1):155, 2000.

Victoria Groom, Jimmy Chen, Theresa Johnson, F Arda Kara, and Clifford Nass. Critic, compatriot, or chump?: Responses to robot blame attribution. In *2010 5th ACM/IEEE International Conference on Human-Robot Interaction (HRI)*, pages 211–217. IEEE, 2010.

André Groß, Christian Christian Schütze, Mara Brandt, Britta Wrede, and Birte Richter. RISE: an open-source architecture for interdisciplinary and reproducible human–robot interaction research. *Frontiers in Robotics and AI*, 10:1245501, 2023.

Hatice Gunes, Frank Broz, Chris S Crawford, Astrid Rosenthal-von der Pütten, Megan Strait, and Laurel Riek. Reproducibility in human-robot interaction: Furthering the science of HRI. *Current Robotics Reports*, 3(4): 281–292, 2022.

Amir Haddadi, Elizabeth A. Croft, Brian T. Gleeson, Karon MacLean, and Javier Alcazar. Analysis of task-based gestures in human-robot interaction. In *2013 IEEE International Conference on Robotics and Automation*, pages 2146–2152. IEEE. 2013. doi: 10.1109/ICRA.2013.6630865.

Sami Haddadin. *Towards Safe Robots: Approaching Asimov's 1st Law*, volume 90, Springer, 2013.

Foad Hamidi, Morgan Klaus Scheuerman, and Stacy M Branham. Gender recognition or gender reductionism? The social implications of embedded gender recognition systems. In *Proceedings of the 2018 Chi Conference on Human Factors in Computing Systems*, pages 1–13. Association for Computing Machinery, 2018.

Jared Hamilton, Nhan Tran, and Tom Williams. Tradeoffs between effectiveness and social perception when using mixed reality to supplement gesturally limited robots. In *Proceedings of the 3rd International Workshop on Virtual, Augmented, and Mixed Reality for HRI*, Association for Computing Machinery (ACM), 2020.

PA Hancock, Theresa T Kessler, Alexandra D Kaplan, John C Brill, and James L Szalma. Evolving trust in robots: specification through sequential and comparative meta-analyses. *Human Factors*, 63(7):1196–1229, 2020. https://doi.org/10.1177/0018720820922080

Nicolai Brodersen Hansen, Christian Dindler, Kim Halskov, Ole Sejer Iversen, Claus Bossen, Ditte Amund Basballe, and Ben Schouten. How participatory design works: mechanisms and effects. In *Proceedings of the 31st Australian Conference on Human-Computer-Interaction*, pages 30–41. Association for Computing Machinery, 2019.

Cindy Harmon-Jones, Brock Bastian, and Eddie Harmon-Jones. The discrete emotions questionnaire: A new tool for measuring state self-reported emotions. *PLoS One*, 11(8), 2016.

Christina Harrington and Tawanna R Dillahunt. Eliciting tech futures among black young adults: A case study of remote speculative co-design. In *Proceedings of the 2021 CHI Conference on Human Factors in Computing Systems*, pages 1–15. Association for Computing Machinery, 2021.

Caroline E Harriott, Tao Zhang, Glenna L Buford, and Julie A Adams. Measuring human workload in a collaborative human-robot team. *Journal of Human-Robot Interaction*, 4(2):61–96, 2015.

Christopher G Harris. Dirty deeds done dirt cheap: a darker side to crowdsourcing. In *2011 IEEE Third International Conference on Privacy, Security, Risk and Trust and 2011 IEEE Third International Conference on Social Computing*, pages 1314–1317. IEEE, 2011.

John Harris and Ehud Sharlin. Exploring the affect of abstract motion in social human-robot interaction. In *2011 Ro-Man*, pages 441–448. IEEE, 2011.

Michael R Harwell, Elaine N Rubinstein, William S Hayes, and Corley C Olds. Summarizing Monte Carlo results in methodological research: The one-and two-factor fixed effects ANOVA cases. *Journal of Educational Statistics*, 17 (4):315–339, 1992.

Cathrine Hasse, Stine Trentemøller, and Jessica Sorenson. Special issue on ethnography in human-robot interaction research. *Paladyn, Journal of Behavioral Robotics*, 10(1):180–181, 2019.

David J Hauser and Norbert Schwarz. Attentive turkers: Mturk participants perform better on online attention checks than do subject pool participants. *Behavior Research Methods*, 48(1):400–407, 2016.

Jerry Zhi-Yang He, Zackory Erickson, Daniel S Brown, Aditi Raghunathan, and Anca Dragan. Learning representations that enable generalization in assistive tasks. In *Conference on Robot Learning*, pages 2105–2114. PMLR, 2023.

Megan L Head, Luke Holman, Rob Lanfear, Andrew T Kahn, and Michael D Jennions. The extent and consequences of p-hacking in science. *PLoS Biology*, 13(3):e1002106, 2015.

Roberta Heale and Alison Twycross. Validity and reliability in quantitative studies. *Evidence-Based Nursing*, 18(3):66–67, 2015.

Marcel Heerink, Ben Krose, Vanessa Evers, and Bob Wielinga. Measuring acceptance of an assistive social robot: a suggested toolkit. In *RO-MAN 2009-The 18th IEEE International Symposium on Robot and Human Interactive Communication*, pages 528–533. IEEE, 2009.

Judith S Heinisch, Jérôme Kirchhoff, Philip Busch, Janine Wendt, Oskar von Stryk, and Klaus David. Physiological data for affective computing in HRI with anthropomorphic service robots: The AFFECT-HRI data set. *Scientific Data*, 11(1):333, 2024.

Zachary Henkel, Kenna Baugus, Cindy L Bethel, and David C May. User expectations of privacy in robot assisted therapy. *Paladyn, Journal of Behavioral Robotics*, 10(1):140–159, 2019.

Joseph Henrich, Steven J Heine, and Ara Norenzayan. Most people are not weird. *Nature*, 466(7302):29–29, 2010.

Trevor Higgins. Randomization in clinical trials: Theory and practice. *Current Medical Research and Opinion*, 19(1):67, 2003. http://search.proquest.com/docview/207967010/.

Anthony Hilton and Richard A Armstrong. Statnote 6: Post-hoc ANOVA tests. *Microbiologist*, 2006:34–36, 2006.

Sebastian Höfer, Kostas Bekris, Ankur Handa, Juan Camilo Gamboa, Melissa Mozifian, Florian Golemo, Chris Atkeson, Dieter Fox, Ken Goldberg, John Leonard, et al. Sim2real in robotics and automation: Applications and challenges. *IEEE Transactions on Automation Science and Engineering*, 18(2):398–400, 2021.

Guy Hoffman. Anki, Jibo, and Kuri: What we can learn from social robots that didn't make it. *IEEE Spectrum*, 2019a.

Guy Hoffman. Evaluating fluency in human–robot collaboration. *IEEE Transactions on Human-Machine Systems*, 49(3):209–218, 2019b.

Guy Hoffman and Cynthia Breazeal. Effects of anticipatory action on human-robot teamwork efficiency, fluency, and perception of team. In *Proceedings of the ACM/IEEE International Conference on Human-Robot Interaction*, pages 1–8. Association for Computing Machinery, 2007.

Guy Hoffman and Xuan Zhao. A primer for conducting experiments in human–robot interaction. *ACM Transactions on Human-Robot Interaction (THRI)*, 10(1):1–31, 2020.

Elizabeth A Hoffmann. Open-ended interviews, power, and emotional labor. *Journal of Contemporary Ethnography*, 36(3):318–346, 2007.

Jess Hohenstein and Malte Jung. Ai as a moral crumple zone: The effects of ai-mediated communication on attribution and trust. *Computers in Human Behavior*, 106:106190, 2020.

Immy Holloway and Kathleen Galvin. *Qualitative Research in Nursing and Healthcare*. John Wiley & Sons, 2016.

Susan R Homack. Understanding what ANOVA post hoc tests are, really, 2001. https://eric.ed.gov/?id=ED449222.

Nadine Homburg. How to include humanoid robots into experimental research: a multi-step approach. In *Proceedings of the 51st Hawaii International Conference on System Sciences*, 2018.

Shanee Honig and Tal Oron-Gilad. Understanding and resolving failures in human-robot interaction: Literature review and model development. *Frontiers in Psychology*, 9:861, 2018.

Shanee Honig and Tal Oron-Gilad. Comparing laboratory user studies and video-enhanced web surveys for eliciting user gestures in human-robot interactions. In *Companion of the 2020 ACM/IEEE International Conference on Human-Robot Interaction*, pages 248–250. Association for Computing Machinery, 2020.

Kasper Hornbæk. Some whys and hows of experiments in human–computer interaction. *Foundations and Trends in Human-Computer Interaction*, 5 (4):299–373, 2013.

Aike C Horstmann and Nicole C Krämer. Great expectations? Relation of previous experiences with social robots in real life or in the media and expectancies based on qualitative and quantitative assessment. *Frontiers in Psychology*, 10:939, 2019.

Asbjørn Hróbjartsson, Julie Pildal, An-Wen Chan, Mette T Haahr, Douglas G Altman, and Peter C Gøtzsche. Reporting on blinding in trial protocols and corresponding publications was often inadequate but rarely contradictory. *Journal of Clinical Epidemiology*, 62(9):967–973, 2009.

Hsiu-Fang Hsieh and Sarah E Shannon. Three approaches to qualitative content analysis. *Qualitative Health Research*, 15(9):1277–1288, 2005.

Jason Hsu. *Multiple Comparisons: Theory and Methods*. CRC Press, 1996.

Yanpei Huang, Etienne Burdet, Lin Cao, Phuoc Thien Phan, Anthony Meng Huat Tiong, Pai Zheng, and Soo Jay Phee. Performance evaluation of a foot interface to operate a robot arm. *IEEE Robotics and Automation Letters*, 4(4):3302–3309, 2019.

Yanpei Huang, Etienne Burdet, Lin Cao, Phuoc Thien Phan, Anthony Meng Huat Tiong, and Soo Jay Phee. A subject-specific four-degree-of-freedom foot interface to control a surgical robot. *IEEE/ASME Transactions on Mechatronics*, 25(2):951–963, 2020.

Yanpei Huang, Wenjie Lai, Lin Cao, Etienne Burdet, and Soo Jay Phee. Design and evaluation of a foot-controlled robotic system for endoscopic surgery. *IEEE Robotics and Automation Letters*, 6(2):2469–2476, 2021.

Andreas Huber, Astrid Weiss, and Marjo Rauhala. The ethical risk of attachment how to identify, investigate and predict potential ethical risks in the development of social companion robots. In *2016 11th ACM/IEEE International Conference on Human-Robot Interaction (HRI)*, pages 367–374. IEEE, 2016.

Michael Hyland. *The Nature of Hypothetical Constructs*, pages 42–58. Macmillan Education UK, 1981.

IEEE. IEEE standard model process for addressing ethical concerns during system design. *IEEE Std 7000-2021*, pages 1–82, 2021. doi: 10.1109/IEEESTD.2021.9536679.

Seigo Inoue, Naoki Mori, Masahiro Tsujikawa, Ryota Ishii, Kanjiro Suzuki, Kunitsugu Kondo, and Michiyuki Kawakami. Determinants of step-through gait pattern acquisition in subacute stroke patients. *Progress in Rehabilitation Medicine*, 7, 2022. doi: 10.2490/ prm.20220035.

Bahar Irfan, James Kennedy, Séverin Lemaignan, Fotios Papadopoulos, Emmanuel Senft, and Tony Belpaeme. Social psychology and human-robot interaction: An uneasy marriage. In *Companion of the 2018 ACM/IEEE International Conference on Human-Robot Interaction*, pages 13–20. Association for Computing Machinery, 2018.

Sarah Irwin. Qualitative secondary data analysis: Ethics, epistemology and context. *Progress in Development Studies*, 13(4):295–306, 2013.

ISO. Part 210: Human-centred design for interactive systems. *ISO/TC 159/SC 4 Ergonomics of Human-System Interaction*, 2019.

ISO. ISO/IEC TR 24368:2022 information technology - artificial intelligence - overview of ethical and societal concerns. https://www.iso.org/standard/78507.html, 2022. Accessed: 2025-03-12.

Takamune Izui and Gentiane Venture. Correlation analysis for predictive models of robot user's impression: A study on visual medium and mechanical noise. *International Journal of Social Robotics*, 12:425–439, 2020. doi: 10.1007/s12369-019-00601-3.

Ryan Blake Jackson, Tom Williams, and Nicole Smith. Exploring the role of gender in perceptions of robotic noncompliance. In *Proceedings of the 2020 ACM/IEEE International Conference on Human-Robot Interaction*, pages 559–567. IEEE, 2020.

An Jacobs, Shirley A Elprama, and Charlotte IC Jewell. Evaluating human-robot interaction with ethnography. In *Human-Robot Interaction*, pages 269–286. Springer, 2020.

Jiun-Yin Jian, Ann M Bisantz, and Colin G Drury. Foundations for an empirically determined scale of trust in automated systems. *International Journal of Cognitive Ergonomics*, 4(1):53–71, 2000.

Shu Jiang and Ronald C. Arkin. Mixed-initiative human-robot interaction: Definition, taxonomy, and survey. In *2015 IEEE International Conference on Systems, Man, and Cybernetics*, pages 954–961, 2015. doi: 10.1109/SMC.2015.174.

Wafa Johal, Séverin Lemaignan, Dražen Bršćić, Marynel Vazquez, and Vicky Charisi. HRI 2025 conference full paper submission guidelines. https://humanrobotinteraction.org/2025/full-papers/, 2024. Accessed: 2024-10-23.

Robert Burke Johnson, Larry B. Christensen, and Burke Johnson. *Educational Research: Quantitative, Qualitative, and Mixed Approaches*. Sage Publications, Inc, 2016.

James M Johnston and Henry S Pennypacker. *Strategies and Tactics of Behavioral Research*. Routledge, 2010.

Kimberlee Jordan, Marcus King, Sophia Hellersteth, Anna Wirén, and Hilda Mulligan. Feasibility of using a humanoid robot for enhancing attention and social skills in adolescents with autism spectrum disorder. *International Journal of Rehabilitation Research*, 36(3):221–227, 2013.

Céline Jost, Brigitte Le Pévédic, Tony Belpaeme, Cindy Bethel, Dimitrios Chrysostomou, Nigel Crook, Marine Grandgeorge, and Nicole Mirnig. *Human-Robot Interaction: Evaluation Methods and Their Standardization*, volume 12, Springer Nature, 2020.

Malte Jung and Pamela Hinds. Robots in the wild: A time for more robust theories of human-robot interaction. *ACM Transactions on Human-Robot Interaction (THRI)*, 7(1):1–5, 2018.

Takayuki Kanda, Masahiro Shiomi, Zenta Miyashita, Hiroshi Ishiguro, and Norihiro Hagita. An affective guide robot in a shopping mall. In *Proceedings of the 4th ACM/IEEE International Conference on Human Robot Interaction*, pages 173–180. IEEE, 2009.

Minsoo Kang, Brian G Ragan, and Jae-Hyeon Park. Issues in outcomes research: an overview of randomization techniques for clinical trials. *Journal of Athletic Training*, 43(2):215–221, 2008a.

Minsoo Kang, Brian G. Ragan, and Jae-Hyeon Park. Issues in outcomes research: An overview of randomization techniques for clinical trials. *Journal of Athletic Training*, 43(2):215–221, 2008b. doi: 10.4085/1062-6050-43.2.215.

Immanuel Kant. *Moral Law: Groundwork of the Metaphysics of Morals.* Routledge, 2013.

Alexandra D Kaplan, Tracy Sanders, and Peter A Hancock. The relationship between extroversion and the tendency to anthropomorphize robots: A Bayesian analysis. *Frontiers in Robotics and AI*, 5:135, 2019.

Maurits Kaptein and Judy Robertson. Rethinking statistical analysis methods for CHI. In *Proceedings of the SIGCHI Conference on Human Factors in Computing Systems*, pages 1105–1114. Association for Computing Machinery, 2012.

John F Kelley. An iterative design methodology for user-friendly natural language office information applications. *ACM Transactions on Information Systems (TOIS)*, 2(1):26–41, 1984.

Janet Kelly. Towards ethical principles for participatory design practice. *CoDesign*, 15(4):329–344, 2019.

Riko Kelter. Bayesian alternatives to null hypothesis significance testing in biomedical research: a non-technical introduction to Bayesian inference with JASP. *BMC Medical Research Methodology*, 20, 2020.

D. Casey Kerrigan, Elizabeth P. Frates, Shannon Rogan, and Patrick O. Riley. Hip hiking and circumduction: Quantitative definitions. *American Journal of Physical Medicine & Rehabilitation*, 79(3):247–252, 2000. doi: 10.1097/00002060-200005000-00006.

Sara Khan and Claudio Germak. Reframing HRI design opportunities for social robots: Lessons learnt from a service robotics case study approach using UX for HRI. *Future Internet*, 10(10):101, 2018.

Michelle E Kiger and Lara Varpio. Thematic analysis of qualitative data: Amee guide no. 131. *Medical Teacher*, 42(8):846–854, 2020.

Callie Y Kim, Christine P Lee, and Bilge Mutlu. Understanding large-language model (LLM)-powered human-robot interaction. In *Proceedings of the 2024 ACM/IEEE International Conference on Human-Robot Interaction*, Association for Computing Machinery (ACM), pages 371–380, 2024.

Dong-Hyun Kim, Yong-Guk Go, and Soo-Mi Choi. An aerial mixed-reality environment for first-person-view drone flying. *Applied Sciences*, 10(16):5436, 2020.

Sung-Hee Kim, Ji Soo Yi, and Niklas Elmqvist. Oopsy-daisy: failure stories in quantitative evaluation studies for visualizations. In *Proceedings of the Fifth Workshop on Beyond Time and Errors: Novel Evaluation Methods for Visualization*, pages 142–146. Association for Computing Machinery, 2014.

Hiroki Kitagawa, Toshihito Nomura, Tanuza Nazmul, Keitaro Omori, Norifumi Shigemoto, Takemasa Sakaguchi, and Hiroki Ohge. Effectiveness of 222-nm ultraviolet light on disinfecting SARS-CoV-2 surface contamination. *American Journal of Infection Control*, 49(3):299–301, 2021.

Barbara Kitchenham, Lech Madeyski, David Budgen, Jacky Keung, Pearl Brereton, Stuart Charters, Shirley Gibbs, and Amnart Pohthong. Robust statistical methods for empirical software engineering. *Empirical Software Engineering*, 22(2):579–630, 2017.

Phyllis Knaack. Phenomenological research. *Western Journal of Nursing Research*, 6(1):107–114, 1984.

Heather Knight, Deanna Flynn, Theing Mwe Oo, and Julia Hansen. Iterative robot waiter algorithm design: Service expectations and social factors. In *Proceedings of the 2024 ACM/IEEE International Conference on Human-Robot Interaction*, pages 394–402. IEEE, 2024.

Kheng Lee Koay, Dag Sverre Syrdal, Michael L Walters, and Kerstin Dautenhahn. Living with robots: Investigating the habituation effect in participants' preferences during a longitudinal human-robot interaction study. In *RO-MAN 2007 - The 16th IEEE International Symposium on Robot and Human Interactive Communication*, pages 564–569. IEEE, 2007. doi: 10.1109/ROMAN.2007.4415149.

Kheng Lee Koay, Dag Sverre Syrdal, Michael L Walters, and Kerstin Dautenhahn. Five weeks in the robot house – exploratory human-robot interaction trials in a domestic setting. In *2009 Second International Conferences on Advances in Computer-Human Interactions*, pages 219–226. IEEE, 2009.

Sharon M Kolb. Grounded theory and the constant comparative method: Valid research strategies for educators. *Journal of Emerging Trends in Educational Research and Policy Studies*, 3(1):83–86, 2012.

Terry K Koo and Mae Y Li. A guideline of selecting and reporting intraclass correlation coefficients for reliability research. *Journal of Chiropractic Medicine*, 15(2):155–163, 2016.

Stefan Kopp, Brigitte Krenn, Stacy Marsella, Andrew N Marshall, Catherine Pelachaud, Hannes Pirker, Kristinn R Thórisson, and Hannes Vilhjálmsson. Towards a common framework for multimodal generation: The behavior markup language. In *Intelligent Virtual Agents: 6th International Conference, IVA 2006, Marina Del Rey, CA, USA, August 21-23, 2006. Proceedings 6*, pages 205–217. Springer, 2006.

Stacy Kowalczyk and Kalpana Shankar. Data sharing in the sciences. *Annual Review of Information Science and Technology*, 45(1):247–294, 2011.

Helena Chmura Kraemer, Jim Mintz, Art Noda, Jared Tinklenberg, and Jerome A. Yesavage. Caution Regarding the Use of Pilot Studies to Guide Power Calculations for Study Proposals. *Archives of General Psychiatry*, 63(5):484–489, 2006. doi: 10.1001/archpsyc.63.5.484.

Stephanie A Kraft, Mildred K Cho, Katherine Gillespie, Meghan Halley, Nina Varsava, Kelly E Ormond, Harold S Luft, Benjamin S Wilfond, and Sandra Soo-Jin Lee. Beyond consent: building trusting relationships with diverse populations in precision medicine research. *The American Journal of Bioethics*, 18(4):3–20, 2018.

Joachim Krauth. *Experimental Design: A Handbook and Dictionary for Medical and Behavioral Research*, volume 14, Elsevier, 2000.

Margaret Mary Krupp, Matthew Rueben, Cindy M Grimm, and William D Smart. Privacy and telepresence robotics: what do non-scientists think? In *Proceedings of the Companion of the 2017 ACM/IEEE International Conference on Human-Robot Interaction*, pages 175–176. Association for Computing Machinery, 2017.

John K Kruschke and Torrin M Liddell. The Bayesian new statistics: Hypothesis testing, estimation, meta-analysis, and power analysis from a Bayesian perspective. *Psychonomic Bulletin & Review*, 25(1):178–206, 2018.

Dana Kulić and Elizabeth A Croft. Affective state estimation for human–robot interaction. *IEEE Transactions on Robotics*, 23(5):991–1000, 2007.

Dana Kulić, Gentiane Venture, Katsu Yamane, Emel Demircan, Ikuo Mizuuchi, and Katja Mombaur. Anthropomorphic movement analysis and synthesis: A survey of methods and applications. *IEEE Transactions on Robotics*, 32(4):776–795, 2016. doi: 10.1109/TRO.2016.2587744.

Guy Laban, Jean-Noël George, Val Morrison, and Emily S. Cross. Tell me more! Assessing interactions with social robots from speech. *Paladyn, Journal of Behavioral Robotics*, 12(1):136–159, 2021.

Daniël Lakens. Calculating and reporting effect sizes to facilitate cumulative science: a practical primer for t-tests and ANOVAs. *Frontiers in Psychology*, 4:863, 2013.

Justin D Lane and David L Gast. Visual analysis in single case experimental design studies: Brief review and guidelines. *Neuropsychological Rehabilitation*, 24(3–4):445–463, 2014.

Kyle M. Lang, Shauna J. Sweet, and Elizabeth M. Grandfield. Getting beyond the null: Statistical modeling as an alternative framework for inference in developmental science. *Research in Human Development*, 14(4):287–304, 2017. doi: 10.1080/15427609.2017.1371567.

Ellen J Langer and Judith Rodin. The effects of choice and enhanced personal responsibility for the aged: a field experiment in an institutional setting. *Journal of Personality and Social Psychology*, 34(2):191, 1976.

Przemyslaw A. Lasota, Terrence Fong, and Julie A. Shah. A survey of methods for safe human-robot interaction. *Foundations and Trends® in Robotics*, 5(4):261–349, 2017. doi: 10.1561/2300000052.

L.E. Lassiter. *The Chicago Guide to Collaborative Ethnography*. Chicago GDES Writing Editing and Publishing CGWEP (CHUP) Series. University of Chicago Press, 2005. https://books. google.com.au/books?id=LItTwAAdtjMC.

Theresa Law, Meia Chita-Tegmark, and Matthias Scheutz. The interplay between emotional intelligence, trust, and gender in human-robot interaction. *International Journal of Social Robotics*, 13(2):297–309, 2021.

Jonathan Lazar, Jinjuan Heidi Feng, and Harry Hochheiser. *Research Methods in Human-Computer Interaction*. Morgan Kaufmann, 2017.

Susan Leavy. Gender bias in artificial intelligence: The need for diversity and gender theory in machine learning. In *Proceedings of the 1st International Workshop on Gender Equality in Software Engineering*, pages 14–16. Association for Computing Machinery, 2018.

Alison Ledgerwood, Courtney K Soderberg, and Jehan Sparks. Designing a study to maximize informational value. In Matthew C Makel and Jonathan A Plucker, editors, *Toward a More Perfect Psychology: Improving Trust, Accuracy, and Transparency in research*, pages 33–58. American Psychological Association, 2017.

Hee Rin Lee and Selma Šabanović. Culturally variable preferences for robot design and use in South Korea, Turkey, and the United States. In *Proceedings of the 2014 ACM/IEEE International Conference on Human-Robot Interaction*, pages 17–24, 2014.

Hee Rin Lee, Haodan Tan, and Selma Šabanović. That robot is not for me: Addressing stereotypes of aging in assistive robot design. In *2016 25th IEEE International Symposium on Robot and Human Interactive Communication (RO-MAN)*, pages 312–317. IEEE, 2016.

Hee Rin Lee, Selma Šabanović, Wan-Ling Chang, Shinichi Nagata, Jennifer Piatt, Casey Bennett, and David Hakken. Steps toward participatory design of social robots: mutual learning with older adults with depression. In *Proceedings of the 2017 ACM/IEEE International Conference on Human-Robot Interaction*, pages 244–253. IEEE, 2017.

Hee Rin Lee, EunJeong Cheon, Chaeyun Lim, and Kerstin Fischer. Configuring humans: What roles humans play in HRI research. In *Proceedings of the 2022 ACM/IEEE International Conference on Human-Robot Interaction*, pages 478–492. IEEE, 2022.

Min Kyung Lee, Jodi Forlizzi, Sara Kiesler, Paul Rybski, John Antanitis, and Sarun Savetsila. Personalization in HRI: A longitudinal field experiment. In *2012 7th ACM/IEEE International Conference on Human-Robot Interaction (HRI)*, pages 319–326. IEEE, 2012.

SS-J Lee. The ethical implications of stratifying by race in pharmacogenomics. *Clinical Pharmacology & Therapeutics*, 81(1):122–125, 2007.

Nancy L. Leech and Anthony J. Onwuegbuzie. An array of qualitative data analysis tools: A call for data analysis triangulation. *School Psychology Quarterly*, 22:557–584, 2007.

Nancy L Leech and Anthony J Onwuegbuzie. Qualitative data analysis: A compendium of techniques and a framework for selection for school psychology research and beyond. *School Psychology Quarterly*, 23(4):587, 2008.

Nancy L Leech and Anthony J Onwuegbuzie. A typology of mixed methods research designs. *Quality & Quantity*, 43(2):265–275, 2009.

Benedikt Leichtmann and Verena Nitsch. How much distance do humans keep toward robots? Literature review, meta-analysis, and theoretical considerations on personal space in human-robot interaction. *Journal of Environmental Psychology*, 68:101386, 2020. doi: 10.1016/j.jenvp.2019.101386.

Séverin Lemaignan, Alexis Jacq, Deanna Hood, Fernando Garcia, Ana Paiva, and Pierre Dillenbourg. Learning by teaching a robot: The case of handwriting. *IEEE Robotics & Automation Magazine*, 23(2):56–66, 2016.

Bruno Leme, Chiara Piezzo, Masakazu Hirokawa, and Kenji Suzuki. Design of a cloud-based robotic platform for accompanying and interacting with humans. In *International Conference on Social Robotics*, pages 262–271. Springer, 2017.

Bruno Leme, Masakazu Hirokawa, Hideki Kadone, and Kenji Suzuki. A socially assistive mobile platform for weight-support in gait training. *International Journal of Social Robotics*, 13(3):459–468, 2021.

Trudo Lemmens and Benjamin Freedman. Ethics review for sale? Conflict of interest and commercial research review boards. *The Milbank Quarterly*, 78(4):547–584, 2000.

Frederick TL Leong and James T Austin. *The Psychology Research Handbook: A Guide for Graduate Students and Research Assistants*. Sage, 2006.

Jessica Nina Lester, Yonjoo Cho, and Chad R Lochmiller. Learning to do qualitative data analysis: A starting point. *Human Resource Development Review*, 19(1):94–106, 2020.

Velvetina Lim, Maki Rooksby, and Emily S. Cross. Social robots on a global stage: Establishing a role for culture during human-robot interaction. *International Journal of Social Robotics*, 13(6):1307–1333, 2021. doi: 10.1007/s12369-020-00710-4.

Yoong Ping Lim, Yi-Chung Lin, and Marcus G. Pandy. Lower-limb muscle function in healthy young and older adults across a range of walking speeds. *Gait & Posture*, 94:124–130, 2022. doi: 10.1016/j.gaitpost.2022.03.003.

Ana Cristina Lindsay, Madelyne J Valdez, Joanna A Pineda, and Mario A Muñoz. Lessons learned from recruiting latino fathers in child health research. *Health Promotion Practice*, 22(4):462–468, 2021.

Mai Skjott Linneberg and Steffen Korsgaard. Coding qualitative data: A synthesis guiding the novice. *Qualitative Research Journal*, 19:259-270, 2019.

Sebastian Linxen, Christian Sturm, Florian Brühlmann, Vincent Cassau, Klaus Opwis, and Katharina Reinecke. How weird is chi? In *Proceedings of the 2021 CHI Conference on Human Factors in Computing Systems*, pages 1–14. Association for Computing Machinery, 2021.

Todd D. Little, Trent D. Buskirk, Lisa M. Willoughby, and Terry T. Tomazic. Nonparametric statistical techniques. In *The Oxford Handbook of Quantitative Methods in Psychology: Vol. 2: Statistical Analysis*, pages 1–68. Oxford University Press, 2013.

Mark S Litwin and Arlene Fink. *How to Assess and Interpret Survey Psychometrics*, volume 8, Sage, 2003.

Hongyi Liu and Lihui Wang. Gesture recognition for human-robot collaboration: A review. *International Journal of Industrial Ergonomics*, 68:355–367, 2018. doi: 10.1016/j.ergon.2017.02.004.

Zeyi Liu, Arpit Bahety, and Shuran Song. REFLECT: Summarizing robot experiences for failure explanation and correction. In *Conference on Robot Learning*, pages 3468–3484. PMLR, 2023.

Elizabeth F Loftus and JC Palmer. Eyewitness testimony. In *Introducing Psychological Research*, pages 305–309. Springer, 1996.

Maria Luce Lupetti, Cristina Zaga, and Nazli Cila. Designerly ways of knowing in HRI: Broadening the scope of design-oriented HRI through the concept of intermediate-level knowledge. In *Proceedings of the 2021 ACM/IEEE International Conference on Human-Robot Interaction*, pages 389–398. Association for Computing Machinery, 2021.

Gil Luria and Sara Rosenblum. Comparing the handwriting behaviours of true and false writing with computerized handwriting measures. *Applied Cognitive Psychology*, 24(8):1115–1128, 2010.

Alexander Ly, Akash Raj, Alexander Etz, Maarten Marsman, Quentin F Gronau, and Eric-Jan Wagenmakers. Bayesian reanalyses from summary statistics: A guide for academic consumers. *Advances in Methods and Practices in Psychological Science*, 1(3):367–374, 2018.

Steve Macenski, Alberto Soragna, Michael Carroll, and Zhenpeng Ge. Impact of ROS 2 node composition in robotic systems. *IEEE Robotics and Automation Letters*, 8(7):3996–4003, 2023.

Steven Macenski, Tully Foote, Brian Gerkey, Chris Lalancette, and William Woodall. Robot Operating System 2: Design, architecture, and uses in the wild. *Science Robotics*, 7(66):eabm6074, 2022.

Alasdair MacIntyre. *Whose justice? Which rationality?* Routledge, 2020.

Kamal Mahtani, Elizabeth A Spencer, Jon Brassey, and Carl Heneghan. Catalogue of bias: observer bias. *BMJ Evidence-Based Medicine*, 23(1):23–24, 2018.

Kirsti Malterud, Volkert Dirk Siersma, and Ann Dorrit Guassora. Sample size in qualitative interview studies: guided by information power. *Qualitative Health Research*, 26(13):1753–1760, 2016.

John C Mankins et al. Technology readiness levels. *White Paper, April*, 6 1995.

Gustav Markkula and Mehmet Dogar. Models of human behavior for human-robot interaction and automated driving: How accurate do the models of human behavior need to be? *IEEE Robotics & Automation Magazine*, 31(3):115–120, 2024. doi: 10.1109/MRA.2022.3182892.

Jeremy A Marvel, Shelly Bagchi, Megan Zimmerman, and Brian Antonishek. Towards effective interface designs for collaborative HRI in manufacturing: Metrics and measures. *ACM Transactions on Human-Robot Interaction (THRI)*, 9(4):1–55, 2020.

Mark Mason. Sample size and saturation in PhD studies using qualitative interviews. *Forum Qualitative Sozialforschung/Forum: Qualitative Social Research*, 11(3), 2010.

Kory W. Mathewson and Patrick M. Pilarski. A brief guide to designing and evaluating human-centered interactive machine learning, 2022. URL https://arxiv.org/abs/2204.09622.

John W Mauchly. Significance test for sphericity of a normal n-variate distribution. *The Annals of Mathematical Statistics*, 11(2):204–209, 1940.

Scott E Maxwell, Ken Kelley, and Joseph R Rausch. Sample size planning for statistical power and accuracy in parameter estimation. *Annual Review of Psychology*, 59, 2008.

Scott E Maxwell, Michael Y Lau, and George S Howard. Is psychology suffering from a replication crisis? What does "failure to replicate" really mean? *American Psychologist*, 70(6):487, 2015.

John H McDonald. *Handbook of biological statistics*, volume 2, Sparky house publishing, 2009.

Mary L McHugh. Interrater reliability: The kappa statistic. *Biochemia Medica*, 22(3):276–282, 2012.

Antonia Meissner, Angelika Trübswetter, Antonia S. Conti-Kufner, and Jonas Schmidtler. Friend or foe? Understanding assembly workers' acceptance of human-robot collaboration. *Journal - Human-Robot Interaction*, 10(1), 2020.

Christine Benedichte Meyer. A case in case study methodology. *Field Methods*, 13(4):329–352, 2001.

John Stuart Mill and Jeremy Bentham. *Utilitarianism and Other Essays.* Penguin, 1987.

Jane Mills and Melanie Birks. *Qualitative Methodology: A Practical Guide.* Sage, 2014.

Michael Mintrom, Shanti Sumartojo, Dana Kulić, Leimin Tian, Pamela Carreno-Medrano, and Aimee Allen. Robots in public spaces: Implications for policy design. *Policy Design and Practice*, 5(2):123–139, 2022.

Nicole Mirnig, Gerald Stollnberger, Markus Miksch, Susanne Stadler, Manuel Giuliani, and Manfred Tscheligi. To err is robot: How humans assess and act toward an erroneous social robot. *Frontiers in Robotics and AI*, 4:21, 2017.

AJung Moon, Daniel M Troniak, Brian Gleeson, Matthew KXJ Pan, Minhua Zheng, Benjamin A Blumer, Karon MacLean, and Elizabeth A Croft. Meet me where I'm gazing: how shared attention gaze affects human-robot handover timing. In *Proceedings of the 2014 ACM/IEEE International Conference on Human-Robot Interaction*, pages 334–341. IEEE, 2014.

Charity J Morgan. Use of proper statistical techniques for research studies with small samples. *American Journal of Physiology-Lung Cellular and Molecular Physiology*, 313(5):L873–L877, 2017.

Alana Mothershaw. When and how you should reuse research participants. https://www.simpleusability.com/inspiration/2018/07/when-and-how-you-should-reuse-research-participants/, 2018. Accessed: 2020-07-27.

Wendy Moyle, Cindy J Jones, Jenny E Murfield, Lukman Thalib, Elizabeth RA Beattie, David KH Shum, Siobhan T O'Dwyer, M Cindy Mervin, and Brian M Draper. Use of a robotic seal as a therapeutic tool to improve dementia symptoms: a cluster-randomized controlled trial. *Journal of the American Medical Directors Association*, 18(9):766–773, 2017.

Wendy Moyle, Cindy Jones, Jenny Murfield, Lukman Thalib, Elizabeth Beattie, David Shum, and Brian Draper. Using a therapeutic companion robot for dementia symptoms in long-term care: reflections from a cluster-RCT. *Aging & Mental Health*, 23(3):329–336, 2019.

Bonnie M Muir. Trust in automation: Part i. theoretical issues in the study of trust and human intervention in automated systems. *Ergonomics*, 37(11):1905–1922, 1994.

Vincent C Müller. Ethics of artificial intelligence and robotics. In Edward N. Zalta (ed.), *Stanford Encyclopedia of Philosophy*, page 1-70. Stanford University, 2020.

Jonathan Mumm and Bilge Mutlu. Human-robot proxemics: physical and psychological distancing in human-robot interaction. In *Proceedings of the 6th International Conference on Human-Robot Interaction*, pages 331–338, 2011.

Fabio Muratore, Fabio Ramos, Greg Turk, Wenhao Yu, Michael Gienger, and Jan Peters. Robot learning from randomized simulations: A review. *Frontiers in Robotics and AI*, 9:799893, 2022.

Robin R Murphy and Jennifer L Burke. Up from the rubble: Lessons learned about HRI from search and rescue. *Proceedings of the Human Factors and Ergonomics Society Annual Meeting*, 49(3):437–441, 2005.

Robin R Murphy and Debra Schreckenghost. Survey of metrics for human-robot interaction. In *2013 8th ACM/IEEE International Conference on Human-Robot Interaction (HRI)*, pages 197–198. IEEE, 2013.

Elaine M. Murtagh, Jacqueline L. Mair, Elroy Aguiar, Catrine Tudor-Locke, and Marie H. Murphy. Outdoor walking speeds of apparently healthy adults: A systematic review and meta-analysis. *Sports Medicine*, 51(1):125–141, 2020. doi: 10.1007/s40279-020-01351-3.

Bilge Mutlu and Jodi Forlizzi. Robots in organizations: The role of workflow, social, and environmental factors in human-robot interaction. In *2008 3rd ACM/IEEE International Conference on Human-Robot Interaction (HRI)*, pages 287–294. IEEE, 2008.

Manisha Natarajan and Matthew Gombolay. Effects of anthropomorphism and accountability on trust in human robot interaction. In *Proceedings of the 2020 ACM/IEEE International Conference on Human-Robot Interaction*, pages 33–42. Association for Computing Machinery, 2020.

Lisa Newington and Alison Metcalfe. Factors influencing recruitment to research: qualitative study of the experiences and perceptions of research teams. *BMC Medical Research Methodology*, 14:1–11, 2014.

NHMRC. National statement on ethical conduct in human research 2023. https://www.nhmrc.gov.au/sites/default/files/documents/attachments/publications/National-Statement-Ethical-Conduct-Human-Research-2023.pdf, 2023. Accessed: 2024-08-09.

Kim F Nimon. Statistical assumptions of substantive analyses across the general linear model: a mini-review. *Frontiers in Psychology*, 3:322, 2012.

Tatsuya Nomura. Robots and gender. *Gender and the Genome*, 1(1):18–25, 2017.

Tatsuya Nomura, Tomohiro Suzuki, Takayuki Kanda, and Kensuke Kato. Measurement of negative attitudes toward robots. *Interaction Studies*, 7 (3):437–454, 2006.

Beau Norgeot, Kathleen Muenzen, Thomas A Peterson, Xuancheng Fan, Benjamin S Glicksberg, Gundolf Schenk, Eugenia Rutenberg, Boris Oskotsky, Marina Sirota, Jinoos Yazdany, et al. Protected health information filter (Philter): accurately and securely de-identifying free-text clinical notes. *NPJ Digital Medicine*, 3(1):1–8, 2020.

Domen Novak, Matjaž Mihelj, and Marko Munih. Psychophysiological responses to different levels of cognitive and physical workload in haptic interaction. *Robotica*, 29(3):367–374, 2011.

Lorelli S Nowell, Jill M Norris, Deborah E White, and Nancy J Moules. Thematic analysis: Striving to meet the trustworthiness criteria. *International Journal of Qualitative Methods*, 16(1):1609406917733847, 2017.

Jean Providence Nzabonimpa. Quantitizing and qualitizing (im-)possibilities in mixed methods research. *Methodological Innovations*, 11(2): 2059799118789021, 2018.

OASH. International compilation of human research standards 2021 edition. https://www.hhs.gov/sites/default/files/ohrp-international-compilation-2021.pdf, 2021. Accessed: 2024-08-09.

Sackett Catalogue of Bias Collaboration, EA Spencer, K Mahtani, et al. Hawthorne bias. *Catalogue of Bias*, 2017.

Harley Oliff, Ying Liu, Maneesh Kumar, and Michael Williams. Improving human–robot interaction utilizing learning and intelligence: A human factors-based approach. *IEEE Transactions on Automation Science and Engineering*, 17(3):1597–1610, 2020.

Dan R Olsen and Michael A Goodrich. Metrics for evaluating human-robot interactions. In *Proceedings of PERMIS*, volume 2003, page 4, Association for Computing Machinery (ACM), 2003.

Abby O'Neill, Abdul Rehman, Abhiram Maddukuri, Abhishek Gupta, Abhishek Padalkar, Abraham Lee, Acorn Pooley, Agrim Gupta, Ajay Mandlekar, Ajinkya Jain, et al. Open X-Embodiment: Robotic learning datasets and RT-X models : Open X-Embodiment collaboration0. In *2024 IEEE International Conference on Robotics and Automation (ICRA)*, pages 6892–6903. IEEE, 2024.

Tõnu Viik. Falling in love with robots: a phenomenological study of experiencing technological alterities. *Paladyn, Journal of Behavioral Robotics*, 11(1):52–65, 2020.

Anthony J Onwuegbuzie and John H Hitchcock. Advanced mixed analysis approaches. In *The Oxford Handbook of Multimethod and Mixed Methods Research Inquiry*. Oxford University Press, 2015.

Anthony J Onwuegbuzie and R Burke Johnson. *The Routledge Reviewer's Guide to Mixed Methods Analysis*. Routledge, 2021.

Anthony J. Onwuegbuzie, Wendy B. Dickinson, Nancy L. Leech, and Annmarie G. Zoran. A qualitative framework for collecting and analyzing data in focus group research. *International Journal of Qualitative Methods*, 8(3):1–21, 2009.

Anthony J Onwuegbuzie, Nancy L Leech, and Kathleen MT Collins. Innovative data collection strategies in qualitative research. *Qualitative Report*, 15(3):696–726, 2010.

Anastasia K Ostrowski, Hae Won Park, and Cynthia Breazeal. Design research in HRI: Roboticists, design features, and users as co-designers. In *Workshop on Designerly HRI Knowledge*, 2020.

Anastasia K Ostrowski, Cynthia Breazeal, and Hae Won Park. Long-term co-design guidelines: Empowering older adults as co-designers of social robots. In *2021 30th IEEE International Conference on Robot & Human Interactive Communication (RO-MAN)*, pages 1165–1172. IEEE, 2021.

Sharon Oviatt. Technology as infrastructure for dehumanization: Three hundred million people with the same face. In *Proceedings of the 2021 International Conference on Multimodal Interaction*, pages 278–287. Association for Computing Machinery, 2021.

Maike Paetzel, Giulia Perugia, and Ginevra Castellano. The persistence of first impressions: The effect of repeated interactions on the perception of a social robot. In *Proceedings of the 2020 ACM/IEEE International Conference on Human-Robot Interaction*, pages 73–82. IEEE, 2020.

Stefan Palan and Christian Schitter. Prolific.ac—A subject pool for online experiments. *Journal of Behavioral and Experimental Finance*, 17:22–27, 2018.

Meta AI Research Papers with Code. Papers with code. https://paperswithcode.com/, 2024. Accessed: 2024-09-04.

Raja Parasuraman and Dietrich H Manzey. Complacency and bias in human use of automation: An attentional integration. *Human Factors*, 52(3):381–410, 2010.

Chris A. C. Parker and Elizabeth Croft. J-strips: Haptic joint limit warnings for human-robot interaction. In *Volume 8: Dynamic Systems and Control, Parts A and B*, pages 789–795. ASMEDC, 2010. doi: 10.1115/IMECE2010-40717.

Jaana Parviainen and Jari Pentti Tapio Pirhonen. Vulnerable bodies in human-robot interactions: Embodiment as ethical issue in robot care for the elderly. *Transformations (Wayne)*, 29:104–115, 2017.

Mildred L Patten and Michelle Newhart. *Understanding Research Methods: An Overview of the Essentials*. Taylor & Francis, 2017.

Jonas E. Pedersen, Kristoffer W. Christensen, Damith Herath, and Elizabeth Jochum. I like the way you move: A mixed-methods approach for studying the effects of robot motion on collaborative human robot interaction. In Alan R. Wagner, David Feil-Seifer, Kerstin S. Haring, Silvia Rossi, Thomas Williams, Hongsheng He, and Shuzhi Sam Ge, editors, *Social Robotics*, pages 73–84. Springer International Publishing, 2020.

F.J. Perez-Grau, R. Ragel, F. Caballero, A. Viguria, and A. Ollero. Semi-autonomous teleoperation of UAVs in search and rescue scenarios. In *International Conference on Unmanned Aircraft Systems (ICUAS)*, pages 1066–1074. IEEE, 2017.

Elizabeth Phillips, Xuan Zhao, Daniel Ullman, and Bertram F Malle. What is human-like? Decomposing robots' human-like appearance using the Anthropomorphic roBOT (ABOT) database. In *Proceedings of the 2018 ACM/IEEE International Conference on Human-Robot Interaction*, pages 105–113. IEEE, 2018.

Karola Pitsch, Hideaki Kuzuoka, Yuya Suzuki, Luise Sussenbach, Paul Luff, and Christian Heath. "the first five seconds": Contingent stepwise entry into an interaction as a means to secure sustained engagement in HRI. In *RO-MAN 2009-The 18th IEEE International Symposium on Robot and Human Interactive Communication*, pages 985–991. IEEE, 2009.

Marie Poggenpoel and Chris Myburgh. The researcher as research instrument in educational research: A possible threat to trustworthiness? (A: research instrument). *Education*, 124(2):418–423, 2003.

Elisa Prati, Margherita Peruzzini, Marcello Pellicciari, and Roberto Raffaeli. How to include user eXperience in the design of human-robot interaction. *Robotics and Computer-Integrated Manufacturing*, 68:102072, 2021.

Judith Preissle, Rebecca Glover-Kudon, Elizabeth A Rohan, Jennifer E Boehm, and Amy DeGroff. Putting ethics on the mixed methods map. In *The Oxford Handbook of Multimethod and Mixed Methods Research Inquiry*. Oxford University Press, 2015.

Matthew S Prewett, Kristin N Saboe, Ryan C Johnson, Michael D Coovert, and Linda R Elliott. Workload in human-robot interaction: a review of manipulations and outcomes. *Proceedings of the Human Factors and Ergonomics Society Annual Meeting*, 53(18):1393–1397, 2009.

Matthew S Prewett, Ryan C Johnson, Kristin N Saboe, Linda R Elliott, and Michael D Coovert. Managing workload in human–robot interaction: a review of empirical studies. *Computers in Human Behavior*, 26(5):840–856, 2010.

Paul C Price, Rajiv Jhangiani, I-Chant A Chiang, et al. *Research methods in psychology*. BCCampus, 2015.

Xavier Puig, Eric Undersander, Andrew Szot, Mikael Dallaire Cote, Tsung-Yen Yang, Ruslan Partsey, Ruta Desai, Alexander William Clegg, Michal Hlavac, So Yeon Min, et al. Habitat 3.0: A co-habitat for humans, avatars and robots. *arXiv preprint arXiv:2310.13724*, 2023.

David Puljiz, Gleb Gorbachev, and Björn Hein. Implementation of augmented reality in autonomous warehouses: challenges and opportunities. *arXiv preprint arXiv:1806.00324*, 2018.

Mansour Rahimi and Waldemar Karwowski. A research paradigm in human-robot interaction. *International Journal of Industrial Ergonomics*, 5(1):59–71, 1990. doi: 10.1016/0169-8141(90)90028-Z.

Paul Ralph and Ewan Tempero. Construct validity in software engineering research and software metrics. In *Proceedings of the 22nd International Conference on Evaluation and Assessment in Software Engineering 2018*, pages 13–23. Association for Computing Machinery, 2018.

Kandethody M Ramachandran and Chris P Tsokos. *Mathematical Statistics with Applications in R*. Academic Press, 2020.

Jens Rasmussen. Risk management in a dynamic society: a modelling problem. *Safety Science*, 27(2–3):183–213, 1997.

Pei-Luen Patrick Rau, Ye Li, and Jun Liu. Effects of a social robot's autonomy and group orientation on human decision-making. *Advances in Human-Computer Interaction*, 2013:263721, 2013. doi: 10.1155/2013/263721.

Daniel J Rea, Denise Geiskkovitch, and James E Young. Wizard of awwws: Exploring psychological impact on the researchers in social HRI experiments. In *Proceedings of the Companion of the 2017 ACM/IEEE International Conference on Human-Robot Interaction*, pages 21–29. Association for Computing Machinery, 2017.

James Reason. *Human Error*. Cambridge university press, 1990.

Sherri A. Rehfeld. *The Impact of Mental Transformation Training Across Levels of Automation on Spatial Awareness in Human-robot Interaction*. Doctoral dissertation, University of Central Florida, 2006.

Merle Reimann, Jesper van de Graaf, Nina van Gulik, Stephanie Van De Sanden, Tibert Verhagen, and Koen Hindriks. Social robots in the wild and the novelty effect. In *International Conference on Social Robotics*, pages 38–48. Springer, 2023.

Helen Mary Richards and Lisa Jennifer Schwartz. Ethics of qualitative research: are there special issues for health services research? *Family Practice*, 19(2):135–139, 2002. doi: 10.1093/fampra/19.2.135.

Katie A. Riddoch and Emily. S. Cross. "Hit the robot on the head with this mallet" - Making a case for including more open questions in HRI research. *Frontiers in Robotics and AI*, 8:2, 2021.

Laurel Riek and Don Howard. A code of ethics for the human-robot interaction profession. *Proceedings of WE Robot*, Elsevier, 2014.

Laurel D Riek. Wizard of Oz studies in HRI: a systematic review and new reporting guidelines. *Journal of Human-Robot Interaction*, 1(1):119–136, 2012.

Alexander Rind. Some whys and hows of experiments in human-computer interaction. *Foundations and Trends® in Human-Computer Interaction*, 5(4):299–373, 2011. doi: 10.1561/1100000043.

Ben Robins and Kerstin Dautenhahn. The role of the experimenter in HRI research - a case study evaluation of children with autism interacting with a robotic toy. In *ROMAN 2006-The 15th IEEE International Symposium on Robot and Human Interactive Communication*, pages 646–651. IEEE, 2006.

Nicole Robinson, Jason Williams, David Howard, Brendan Tidd, Fletcher Talbot, Brett Wood, Alex Pitt, Navinda Kottege, and Dana Kulić. Human-robot team performance compared to full robot autonomy in 16 real-world search and rescue missions: Adaptation of the DARPA subterranean challenge. *arXiv preprint arXiv:2212.05626*, 2022.

Nicole L. Robinson and David J. Kavanagh. A social robot to deliver a psychotherapeutic treatment: Qualitative responses by participants in a randomized controlled trial and future design recommendations. *International Journal of Human-Computer Studies*, 155:102700, 2021.

Nicole L Robinson, Jennifer Connolly, Genevieve M Johnson, Yejee Kim, Leanne Hides, and David J Kavanagh. Measures of incentives and confidence in using a social robot. *Science Robotics*, 3(21):eaat6963, 2018.

Nicole L Robinson, Teah-Neal Hicks, Gavin Suddrey, and David J Kavanagh. The robot self-efficacy scale: Robot self-efficacy, likability and willingness to interact increases after a robot-delivered tutorial. In *2020 29th IEEE International Conference on Robot and Human Interactive Communication (RO-MAN)*, pages 272–277. IEEE, 2020.

Nicole Lee Robinson, Timothy Vaughan Cottier, and David John Kavanagh. Psychosocial health interventions by social robots: systematic review of randomized controlled trials. *Journal of MEDICAL INTERNET RESEARCH*, 21(5): e13203, 2019.

Alessandro Roncone, Olivier Mangin, and Brian Scassellati. Transparent role assignment and task allocation in human robot collaboration. In *2017 IEEE International Conference on Robotics and Automation (ICRA)*, pages 1014–1021. IEEE, 2017.

Julia Rosén, Jessica Lindblom, and Erik Billing. Reporting of ethical conduct in human-robot interaction research. In *International Conference on Applied Human Factors and Ergonomics*, pages 87–94. Springer, 2021.

Astrid Rosenthal-von der Pütten, David Sirkin, Anna Abrams, and Laura Platte. The forgotten in HRI: Incidental encounters with robots in public spaces. In *Companion of the 2020 ACM/IEEE International Conference on Human-Robot Interaction*, pages 656–657. Association for Computing Machinery, 2020.

Astrid M Rosenthal-von der Pütten, Nicole C Krämer, Laura Hoffmann, Sabrina Sobieraj, and Sabrina C Eimler. An experimental study on emotional reactions towards a robot. *International Journal of Social Robotics*, 5:17–34, 2013.

Mark A Rothstein. Health privacy in the electronic age. *The Journal of Legal Medicine*, 28(4):487–501, 2007.

Matthew Rueben, Alexander Mois Aroyo, Christoph Lutz, Johannes Schmölz, Pieter Van Cleynenbreugel, Andrea Corti, Siddharth Agrawal, and William D Smart. Themes and research directions in privacy-sensitive robotics. In *2018 IEEE Workshop on Advanced Robotics and its Social Impacts (ARSO)*, pages 77–84. IEEE, 2018.

Matthew Rueben, Shirley A Elprama, Dimitrios Chrysostomou, and An Jacobs. Introduction to (re)using questionnaires in human-robot interaction research. In *Human-Robot Interaction*, pages 125–144. Springer, 2020a.

Matthew Rueben, Eitan Rothberg, and Maja J Matarić. Applying the theory of make-believe to human-robot interaction. In *Culturally Sustainable Social Robotics*, pages 40–50. IOS Press, 2020b.

Matthew Rueben, Jeffrey Klow, Madelyn Duer, Eric Zimmerman, Jennifer Piacentini, Madison Browning, Frank J Bernieri, Cindy M Grimm, and William D Smart. Mental models of a mobile shoe rack: Exploratory findings from a long-term in-the-wild study. *ACM Transactions on Human-Robot Interaction (THRI)*, 10(2):1–36, 2021.

Craig J Russell and Michelle A Dean. To log or not to log: Bootstrap as an alternative to the parametric estimation of moderation effects in the presence of skewed dependent variables. *Organizational Research Methods*, 3(2):166–185, 2000.

John Rust and Susan Golombok. *Modern Psychometrics: The Science of Psychological Assessment*. Routledge, 2014.

G.D. Ruxton. Allocation concealment as a potentially useful aspect of randomised experiments. *Behavioral Ecology and Sociobiology*, 71(2):1–3, 2017.

Inkyu Sa, Stefan Hrabar, and Peter Corke. Inspection of pole-like structures using a visual-inertial aided vtol platform with shared autonomy. *Sensors*, 15:22003–22048, 2015.

Selma Šabanović, Marek P Michalowski, and Reid Simmons. Robots in the wild: Observing human-robot social interaction outside the lab. In *9th IEEE International Workshop on Advanced Motion Control, 2006*, pages 596–601. IEEE, 2006.

Khaled Saleh, Mohammed Hossny, and Saeid Nahavandi. Towards trusted autonomous vehicles from vulnerable road users perspective. In *2017 Annual IEEE International Systems Conference (SysCon)*, pages 1–7. IEEE, 2017.

Maha Salem, Gabriella Lakatos, Farshid Amirabdollahian, and Kerstin Dautenhahn. Would you trust a (faulty) robot? Effects of error, task type and personality on human-robot cooperation and trust. In *2015 10th ACM/IEEE International Conference on Human-Robot Interaction (HRI)*, pages 1–8. IEEE, 2015.

Neil J Salkind. *Encyclopedia of Research Design*, volume 1. Sage, 2010.

Paul M Salmon, Neville A Stanton, Guy H Walker, Daniel Jenkins, Darshna Ladva, Laura Rafferty, and Mark Young. Measuring situation awareness in complex systems: Comparison of measures study. *International Journal of Industrial Ergonomics*, 39(3):490–500, 2009.

Paul M Salmon, Brandon J King, Scott McLean, Gemma JM Read, Christopher Shanahan, and Kate Devitt. A framework of human factors methods for safe, ethical, and usable artificial intelligence in defense. In *Putting AI in the critical loop*, pages 245–263. Elsevier, 2024.

Pericle Salvini, Cecilia Laschi, and Paolo Dario. Design for acceptability: improving robots' coexistence in human society. *International Journal of Social Robotics*, 2(4):451–460, 2010.

Jacob Hernandez Sanchez, Walid Amanhoud, Aude Billard, and Mohamed Bouri. Foot control of a surgical laparoscopic gripper via 5dof haptic robotic platform: Design, dynamics and haptic shared control. In *2021 IEEE International Conference on Robotics and Automation (ICRA)*, pages 12559–12566. IEEE, 2021.

Anara Sandygulova, Wafa Johal, Zhanel Zhexenova, Bolat Tleubayev, Aida Zhanatkyzy, Aizada Turarova, Zhansaule Telisheva, Anna CohenMiller, Thibault Asselborn, and Pierre Dillenbourg. CoWriting Kazakh: Learning a new script with a robot. In *Proceedings of the 2020 ACM/IEEE International Conference on Human-Robot Interaction*, pages 113–120. IEEE, 2020.

Mahnaz Sanjari, Fatemeh Bahramnezhad, Fatemeh Khoshnava Fomani, Mahnaz Shoghi, and Mohammad Ali Cheraghi. Ethical challenges of researchers in qualitative studies: The necessity to develop a specific guideline. *Journal of Medical Ethics and History of Medicine*, 7, 2014.

Vasanth Sarathy, Thomas Arnold, and Matthias Scheutz. When exceptions are the norm: Exploring the role of consent in HRI. *ACM Transactions on Human-Robot Interaction (THRI)*, 8(3):1–21, 2019.

Manolis Savva, Abhishek Kadian, Oleksandr Maksymets, Yili Zhao, Erik Wijmans, Bhavana Jain, Julian Straub, Jia Liu, Vladlen Koltun, Jitendra Malik, et al. Habitat: A platform for embodied AI research. In *Proceedings of the IEEE/CVF International Conference on Computer Vision*, pages 9339–9347. IEEE, 2019.

Brian Scassellati, Henny Admoni, and Maja Matarić. Robots for use in autism research. *Annual Review of Biomedical Engineering*, 14:275–294, 2012.

Daniel J Schad, Michael Betancourt, and Shravan Vasishth. Toward a principled Bayesian workflow in cognitive science. *arXiv preprint arXiv:1904.12765*, 2019.

Kristin E Schaefer. *The perception and measurement of human-robot trust.* PhD thesis, University of Central Florida Orlando, Florida, 2013.

Stefano Scheggi, Fabio Morbidi, and Domenico Prattichizzo. Human-robot formation control via visual and vibrotactile haptic feedback. *IEEE Transactions on Haptics*, 7(4):499–511, 2014. doi: 10.1109/TOH.2014.2332173.

Morgan Klaus Scheuerman, Katta Spiel, Oliver L Haimson, Foad Hamidi, and Stacy M Branham. Hci guidelines for gender equity and inclusivity. *UMBC Faculty Collection*, 2020.

Sebastian Schneider and Franz Kummert. Comparing robot and human guided personalization: adaptive exercise robots are perceived as more competent and trustworthy. *International Journal of Social Robotics*, 13:1–17, 2020.

Jean Scholtz. Theory and evaluation of human robot interactions. In *36th Annual Hawaii International Conference on System Sciences, 2003. Proceedings of the*, pages 10. IEEE, 2003.

Jean Scholtz and Siavosh Bahrami. Human-robot interaction: development of an evaluation methodology for the bystander role of interaction. In *SMC'03 Conference Proceedings. 2003 IEEE International Conference on Systems, Man and Cybernetics. Conference Theme-System Security and Assurance (Cat. No. 03CH37483)*, volume 4, pages 3212–3217. IEEE, 2003.

Judith Schoonenboom and R Burke Johnson. How to construct a mixed methods research design. *KZfSS Kölner Zeitschrift für Soziologie und Sozialpsychologie*, 69(2):107–131, 2017.

Mariah L Schrum, Michael Johnson, Muyleng Ghuy, and Matthew C Gombolay. Four years in review: Statistical practices of Likert scales in human-robot interaction studies. *arXiv preprint arXiv:2001.03231*, 2020.

Kenneth F Schulz and David A Grimes. Allocation concealment in randomised trials: defending against deciphering. *The Lancet*, 359(9306):614–618, 2002.

Russell K. Schutt. Qualitative data analysis. In *Investigating the Social World: the Process and Practice of Research*, Chapter 11. SAGE Publications, 2018.

Yasaman Sadat Sefidgar. *TAMER: touch-guided anxiety management via engagement with a robotic pet efficacy evaluation and the first steps of the interaction design*. PhD thesis, University of British Columbia, 2012. https://open.library.ubc.ca/cIRcle/collections/ubctheses/ 24/items/ 1.0052145.

Johanna Seibt. Towards an ontology of simulated social interaction: Varieties of the "as if" for robots and humans. In Raul Hakli and Johanna Seibt, editors, *Sociality and Normativity for Robots: Philosophical Inquiries into Human-Robot Interactions*, Studies in the Philosophy of Sociality, pages 11–39. Springer International Publishing, 2017. doi: 10.1007/978-3-319-53133-5_2.

Johanna Seibt, Malene Flensborg Damholdt, and Christina Vestergaard. Integrative social robotics, value-driven design, and transdisciplinarity. *Interaction Studies*, 21(1):111–144, 2020.

Johanna Seibt, Christina Vestergaard, and Malene F Damholdt. The complexity of human social interactions calls for mixed methods in HRI: Comment on "a primer for conducting experiments in human-robot interaction" by G. Hoffman and X. Zhao. *ACM Transactions on Human-Robot Interaction (THRI)*, 10(1):1–4, 2021.

Aran Sena and Matthew Howard. Quantifying teaching behavior in robot learning from demonstration. *The International Journal of Robotics Research*, 39(1):54–72, 2020.

Hashini Senaratne, Alex Pitt, Fletcher Talbot, Peyman Moghadam, Pavan Sikka, David Howard, Jason Williams, Dana Kulić, and Cécile Paris. Measuring situational awareness latency in human-robot teaming experiments. In *2023 32nd IEEE International Conference on Robot and Human Interactive Communication (RO-MAN)*, pages 2624–2631. IEEE, 2023.

Mohammad Shaqura, Khalid Alzuhair, Fadl Abdellatif, and Jeff S. Shamma. Human supervised multirotor UAV system design for inspection applications. In *IEEE International Symposium on Safety, Security, and Rescue Robotics (SSRR)*, pages 1–6. IEEE, 2018.

Amanda Sharkey and Noel Sharkey. Granny and the robots: ethical issues in robot care for the elderly. *Ethics and Information Technology*, 14(1):27–40, 2012.

Nicole L. Sharp, Rosalind A. Bye, and Anne Cusick. Narrative analysis. In Pranee Liamputtong, editor, *Handbook of Research Methods in Health Social Sciences*, pages 861–880. Springer, 2019.

David J Sheskin. *Handbook of Parametric and Nonparametric Statistical Procedures*. CRC Press, 2020.

Zhonghao Shi. Awesome LLM for SAR. https://github.com/HaaaO/awesome-LLM-for-social-HRI, 2024. Accessed: 2024-09-04.

Takanori Shibata, Yukitaka Kawaguchi, and Kazuyoshi Wada. Investigation on people living with seal robot at home. *International Journal of Social Robotics*, 4(1):53–63, 2012.

Jaeeun Shim and Ronald C. Arkin. A taxonomy of robot deception and its benefits in HRI. In *2013 IEEE International Conference on Systems, Man, and Cybernetics*, pages 2328–2335. IEEE, 2013. doi: 10.1109/SMC.2013.398.

B. Shneiderman. *Human-Centered AI*. OUP Oxford, 2022. https://books.google.com.au/books?id=mSRXEAAAQBAJ

Elaine Short, Justin Hart, Michelle Vu, and Brian Scassellati. No fair!! An interaction with a cheating robot. In *2010 5th ACM/IEEE International Conference on Human-Robot Interaction (HRI)*, pages 219–226. IEEE, 2010.

Allison Shorten and Joanna Smith. Mixed methods research: expanding the evidence base, 2017.

Patrick E Shrout and Joseph L Rodgers. Psychology, science, and knowledge construction: Broadening perspectives from the replication crisis. *Annual Review of Psychology*, 69:487–510, 2018.

Candace L Sidner, Christopher Lee, Cory D Kidd, Neal Lesh, and Charles Rich. Explorations in engagement for humans and robots. *Artificial Intelligence*, 166(1-2):140–164, 2005.

Joseph P. Simmons, Leif D. Nelson, and Uri Simonsohn. Pre-registration: Why and how. *Journal of Consumer Psychology*, 31:151–162, 2021.

Divyanshu Kumar Singh, Manohar Kumar, Eduard Fosch-Villaronga, Deepa Singh, and Jainendra Shukla. Ethical considerations from child-robot interactions in under-resourced communities. *International Journal of Social Robotics*, 2022. doi: 10.1007/s12369-022-00882-1.

Peter Slattery, Alexander K. Saeri, Emily A. C. Grundy, Jess Graham, Michael Noetel, Risto Uuk, James Dao, Soroush Pour, Stephen Casper, and Neil Thompson. The AI risk repository: A comprehensive meta-review, database, and taxonomy of risks from artificial intelligence. https://airisk.mit.edu/, 2024. Accessed: 2024-08-16.

Jonathan A Smith and Pnina Shinebourne. *Interpretative Phenomenological Analysis*. American Psychological Association, 2012.

Linda J Smith. How ethical is ethical research? Recruiting marginalized, vulnerable groups into health services research. *Journal of Advanced nursing*, 62(2):248–257, 2008.

Nikolai Smolyanskiy and Mar Gonzalez-Franco. Stereoscopic first person view system for drone navigation. *Frontiers in Robotics and AI*, 4:247625, 2017.

Phase Locked Software. Study randomizer. https://www.studyrandomizer.com, 2021.

Jae W Song and Kevin C Chung. Observational studies: cohort and case-control studies. *Plastic and reconstructive surgery*, 126(6):2234, 2010.

Robert Sparrow. Robotics has a race problem. *Science, Technology, & Human Values*, 45(3):538–560, 2020.

Robert Sparrow and Linda Sparrow. In the hands of machines? The future of aged care. *Minds and Machines*, 16(2):141–161, 2006.

Clay Spinuzzi. The methodology of participatory design. *Technical Communication*, 52(2):163–174, 2005.

Micol Spitale, Rebecca Stower, Maria Teresa Parreira, Elmira Yadollahi, Iolanda Leite, and Hatice Gunes. HRI wasnt built in a day: A call to action for responsible HRI research. In *The 33rd IEEE International Conference on Robot and Human Interactive Communication (IEEE RO-MAN 2024)*, 2024. doi: 10.17863/CAM.109265. https://www.repository.cam.ac.uk/handle/1810/369482.

Gerald Steinbauer. A survey about faults of robots used in robocup. In *Robot Soccer World Cup*, pages 344–355. Springer, 2012.

Aaron Steinfeld, Terrence Fong, David Kaber, Michael Lewis, Jean Scholtz, Alan Schultz, and Michael Goodrich. Common metrics for human-robot interaction. In *Proceedings of the 1st ACM SIGCHI/SIGART Conference on Human-Robot Interaction*, HRI '06, pages 33–40. Association for Computing Machinery, 2006. doi: 10.1145/1121241.1121249.

Steve Stemler. An overview of content analysis. *Practical Assessment, Research, and Evaluation*, 7(1):17, 2000.

Megan Strait, Florian Lier, Jasmin Bernotat, Sven Wachsmuth, Friederike Eyssel, Robert Goldstone, and Selma Šabanović. A three-site reproduction of the joint simon effect with the NAO robot. In *Proceedings of the 2020 ACM/IEEE International Conference on Human-Robot Interaction*, pages 103–111, 2020.

Yolande Strengers and Jenny Kennedy. *The Smart Wife: Why Siri, Alexa, and Other Smart Home Devices Need a Feminist Reboot*. MIT Press, 2020.

Roy Suddaby. From the editors: What grounded theory is not. *Academy of Management Journal*, 49(4):633–642, 2006.

Gail M Sullivan and Anthony R Artino Jr. Analyzing and interpreting data from Likert-type scales. *Journal of Graduate Medical Education*, 5(4):541–542, 2013.

Shanti Sumartojo, Robert Lundberg, Leimin Tian, Pamela Carreno-Medrano, Dana Kulić, and Michael Mintrom. Imagining public space robots of the near-future. *Geoforum*, 124:99–109, 2021.

Shanti Sumartojo, Robert Lundberg, Dana Kulić, Leimin Tian, Pamela Carreno-Medrano, Michael Mintrom, Daniele Lugli, and Aimee Allen. The robotic production of spatiality: Predictability, partitioning, and connection. *Transactions of the Institute of British Geographers*, 48(1):56–68, 2023.

Annelie J Sundler, Elisabeth Lindberg, Christina Nilsson, and Lina Palmér. Qualitative thematic analysis based on descriptive phenomenology. *Nursing Open*, 6(3):733–739, 2019.

JaYoung Sung, Henrik I Christensen, and Rebecca E Grinter. Robots in the wild: understanding long-term use. In *Proceedings of the 4th ACM/IEEE International Conference on Human Robot Interaction*, pages 45–52, 2009.

W Newton Suter. *Introduction to educational research: A critical thinking approach*. Sage Publications, 2011.

David M Swanson and Rebecca A Betensky. Research participant compensation: a matter of statistical inference as well as ethics. *Contemporary Clinical Trials*, 45:265–269, 2015.

Daniel Szafir. Mediating human-robot interactions with virtual, augmented, and mixed reality. In *International Conference on Human-Computer Interaction*, pages 124–149. Springer, 2019.

Denes Szucs and John P. A. Ioannidis. When null hypothesis significance testing is unsuitable for research: A reassessment. *Frontiers in Human Neuroscience*, 11:390, 2017.

Roger Andre Søraa, Gunhild Tøndel, Mark W. Kharas, and J. Artur Serrano. What do older adults want from social robots? A qualitative research approach to human-robot interaction (HRI) studies. *International Journal of Social Robotics*, 2022. doi: 10.1007/s12369-022-00914-w.

Barbara G Tabachnick, Linda S Fidell, and Jodie B Ullman. *Using multivariate statistics*, volume 5. Pearson Boston, 2007.

Aaquib Tabrez and Bradley Hayes. Improving human-robot interaction through explainable reinforcement learning. In *2019 14th ACM/IEEE International Conference on Human-Robot Interaction (HRI)*, pages 751–753. IEEE, 2019.

Leila Takayama and Caroline Pantofaru. Influences on proxemic behaviors in human-robot interaction. In *2009 IEEE/RSJ International Conference on Intelligent Robots and Systems*, pages 5495–5502, 2009. doi: 10.1109/IROS.2009.5354145.

Xiang Zhi Tan, Marynel Vázquez, Elizabeth J Carter, Cecilia G Morales, and Aaron Steinfeld. Inducing bystander interventions during robot abuse with social mechanisms. In *Proceedings of the 2018 ACM/IEEE International Conference on Human-Robot Interaction*, pages 169–177, 2018.

Ana Tanevska, Francesco Rea, Giulio Sandini, Lola Cañamero, and Alessandra Sciutti. A socially adaptable framework for human-robot interaction. *arXiv preprint arXiv:2003.11410*, 2020.

Adriana Tapus, Cristian Tapus, and Maja J Mataric. The use of socially assistive robots in the design of intelligent cognitive therapies for people with dementia. In *2009 IEEE International Conference on Rehabilitation Robotics*, pages 924–929. IEEE, 2009.

Kroton Medical Technology. Needle threading task. https://kroton.info/en/laparoscopic/8-needle-threading-lap003/, 2023. [Online; accessed 23-October-2023].

Lehana Thabane, Jinhui Ma, Rong Chu, Ji Cheng, Afisi Ismaila, Lorena P Rios, Reid Robson, Marroon Thabane, Lora Giangregorio, and Charles H Goldsmith. A tutorial on pilot studies: the what, why and how. *BMC Medical Research Methodology*, 10(1):1, 2010.

Kyle A Thomas and Scott Clifford. Validity and Mechanical Turk: An assessment of exclusion methods and interactive experiments. *Computers in Human Behavior*, 77:184–197, 2017.

Andrea Thomaz, Guy Hoffman, and Maya Cakmak. Computational human-robot interaction. *Foundations and Trends in Robotics*, 4(2):105–223, 2016. doi: 10.1561/2300000049.

Bruce Thompson. Common methodology mistakes in dissertations, revisited, 1994. https://eric.ed.gov/?id=ED368771.

Sally Thorne. Data analysis in qualitative research. *Evidence-Based Nursing*, 3(3):68–70, 2000.

Sebastian Thrun. Toward a framework for human-robot interaction. *Human–Computer Interaction*, 19(1-2):9–24, 2004.

Leimin Tian and Sharon Oviatt. A taxonomy of social errors in human-robot interaction. *ACM Transactions on Human-Robot Interaction (THRI)*, 10(2):1–32, 2021.

Leimin Tian, Pamela Carreno-Medrano, Shanti Sumartojo, Michael Mintrom, Enrique Coronado, Gentiane Venture, and Dana Kulić. User expectations of robots in public spaces: A co-design methodology. In *International Conference on Social Robotics*, pages 259–270. Springer, 2020.

Leimin Tian, Pamela Carreno-Medrano, Aimee Allen, Shanti Sumartojo, Michael Mintrom, Enrique Coronado Zuniga, Gentiane Venture, Elizabeth Croft, and Dana Kulić. Redesigning human-robot interaction in response to robot failures: A participatory design methodology. In *Extended Abstracts of the 2021 CHI Conference on Human Factors in Computing Systems*, pages 1–8, 2021.

Leimin Tian, Sharon Oviatt, Michal Muszynski, Brent C. Chamberlain, Jennifer Healey, and Akane Sano. *Addressing Ethical Issues of Affective Computing*, chapter 11. Association for Computing Machinery, 2022.

Lorenza Tiberio, Amedeo Cesta, and Marta Olivetti Belardinelli. Psychophysiological methods to evaluate user's response in human robot interaction: a review and feasibility study. *Robotics*, 2(2):92–121, 2013.

Uthman Tijani, Hong Wang, and Zhao Han. Towards reproducible language-based HRI experiments: Open-sourcing a generalized experiment project. In *Proceedings of the 2024 ACM/IEEE International Conference on Human-Robot Interaction*, pages 949–953, 2024.

Allison Toth, George Banks, David Mellor, Ernest O'Boyle, Ashleigh Dickson, Daniel Davis, Alex DeHaven, Jaime Bochantin, and Jared Borns. Study preregistration: An evaluation of a method for transparent reporting. *Journal of Business and Psychology*, 36, 2021.

Daniel C. Tozadore, Joao P.H. Valentini, Victor H.S. Rodrigues, Fernando M.L. Vendrameto, Rodrigo G. Zavarizz, and Roseli A.F. Romero. Towards adaptation and personalization in task based on human-robot interaction. In *2018 Latin American Robotic Symposium, 2018 Brazilian Symposium on Robotics (SBR) and 2018 Workshop on Robotics in Education (WRE)*, pages 383–389, 2018. doi: 10.1109/LARS/SBR/ WRE.2018.00075.

Margaret L Traeger, Sarah Strohkorb Sebo, Malte Jung, Brian Scassellati, and Nicholas A Christakis. Vulnerable robots positively shape human conversational dynamics in a human–robot team. *Proceedings of the National Academy of Sciences*, 117(12):6370–6375, 2020.

Australian Clinical Trials. Phases of clinical trials. https://www.australianclinicaltrials.gov.au/what-clinical-trial/ phases-clinical-trials, 2015.

Gabriele Trovato, Renato Paredes, Javier Balvin, Francisco Cuellar, Nicolai Bæk Thomsen, Soren Bech, and Zheng-Hua Tan. The sound or silence: investigating the influence of robot noise on proxemics. In *2018 27th IEEE International Symposium on Robot and Human Interactive Communication (RO-MAN)*, pages 713–718. IEEE, 2018.

Nathan Tsoi, Mohamed Hussein, Olivia Fugikawa, JD Zhao, and Marynel Vázquez. An approach to deploy interactive robotic simulators on the web for HRI experiments: Results in social robot navigation. In *2021 IEEE/RSJ International Conference on Intelligent Robots and Systems (IROS)*, pages 7528–7535. IEEE, 2021.

Jos Twisk and Wieke de Vente. Attrition in longitudinal studies: how to deal with missing data. *Journal of Clinical Epidemiology*, 55(4):329–337, 2002.

Daniel Ullman, Salomi Aladia, and Bertram F Malle. Challenges and opportunities for replication science in HRI: A case study in human-robot trust. In *Proceedings of the 2021 ACM/IEEE International Conference on Human-Robot Interaction*, pages 110–118, 2021.

Mojtaba Vaismoradi, Hannele Turunen, and Terese Bondas. Content analysis and thematic analysis: Implications for conducting a qualitative descriptive study. *Nursing & health sciences*, 15(3):398–405, 2013.

Jan Van den Broeck, Solveig Argeseanu Cunningham, Roger Eeckels, and Kobus Herbst. Data cleaning: detecting, diagnosing, and editing data abnormalities. *PLoS Medicine*, 2(10):e267, 2005.

Edwin Van Teijlingen and Vanora Hundley. The importance of pilot studies. *Nursing Standard (through 2013)*, 16(40):33, 2002.

Aimee Van Wynsberghe. Designing robots for care: Care centered value-sensitive design. In *Machine Ethics and Robot Ethics*, pages 185–211. Routledge, 2020.

Joachim Vandekerckhove, Jeffrey N Rouder, and John K Kruschke. Bayesian methods for advancing psychological science. *Psychonomic Bulletin & Review*, 25:1–4, 2018.

Melina C Vassiliou, Brian J Dunkin, Jeffrey M Marks, and Gerald M Fried. FLS and FES: comprehensive models of training and assessment. *The Surgical Clinics of North America*, 90(3):535–558, 2010.

Louise Veling and Conor McGinn. Qualitative research in HRI: A review and taxonomy. *International Journal of Social Robotics*, 13(2):1689–1709, 2021.

Renato Paredes Venero and Alex Davila. Experimental research methodology and statistics insights. In *Human-Robot Interaction*, pages 333–353. Springer, 2020.

Susanne Vogl. Integrating and consolidating data in mixed methods data analysis: Examples from focus group data with children. *Journal of Mixed Methods Research*, 13(4):536–554, 2019.

Paul Vogt, Mirjam De Haas, Chiara De Jong, Peta Baxter, and Emiel Krahmer. Child-robot interactions for second language tutoring to preschool children. *Frontiers in Human Neuroscience*, 11:73, 2017.

Erik Von Elm, Douglas G Altman, Matthias Egger, Stuart J Pocock, Peter C Gøtzsche, and Jan P Vandenbroucke. The strengthening the reporting of observational studies in epidemiology (STROBE) statement: guidelines for reporting observational studies. *Bulletin of the World Health Organization*, 85:867–872, 2007.

Sandrine Voros, Georges-Pascal Haber, Jean-François Menudet, Jean-Alexandre Long, and Philippe Cinquin. Viky robotic scope holder: Initial clinical experience and preliminary results using instrument tracking. *IEEE/ASME Transactions on Mechatronics*, 15(6):879–886, 2010.

Michael L Walters, Kerstin Dautenhahn, Sarah N Woods, and Kheng Lee Koay. Robotic etiquette: results from user studies involving a fetch and carry task. In *2007 2nd ACM/IEEE International Conference on Human-Robot Interaction (HRI)*, pages 317–324. IEEE, 2007.

Ker-Jiun Wang, Mohammad Shidujaman, Caroline Yan Zheng, and Prakash Thakur. HRIpreneur thinking: strategies towards faster innovation and commercialization of academic HRI research. In *2019 IEEE International Conference on Advanced Robotics and its Social Impacts (ARSO)*, pages 219–226. IEEE, 2019.

Lin Wang, Pei-Luen Patrick Rau, Vanessa Evers, Benjamin Krisper Robinson, and Pamela Hinds. When in Rome: the role of culture & context in adherence to robot recommendations. In *2010 5th ACM/IEEE International Conference on Human-Robot Interaction (HRI)*, pages 359–366. IEEE, 2010.

Lun Wang, Luca Iocchi, Andrea Marrella, and Daniele Nardi. HRI users' studies in the context of the SciRoc challenge: Some insights on gender-based differences. In *Proceedings of the 8th International Conference on Human-Agent Interaction*, pages 287–289, 2020.

Y. Wang, F. Guimbretière, and K. E. Green. Are space-making robots, agents? Investigations on user perception of an embedded robotic surface. In *2020 29th IEEE International Conference on Robot and Human Interactive Communication (RO-MAN)*, pages 1230–1235, 2020. doi: 10. 1109/RO-MAN47096.2020.9223532.

Astrid Weiss and Christoph Bartneck. Meta analysis of the usage of the godspeed questionnaire series. In *2015 24th IEEE International Symposium on Robot and Human Interactive Communication (RO-MAN)*, pages 381–388. IEEE, 2015.

Astrid Weiss and Katta Spiel. Robots beyond science fiction: mutual learning in human–robot interaction on the way to participatory approaches. *AI & SOCIETY*, 37(2):501–515, 2022.

Astrid Weiss, Regina Bernhaupt, Michael Lankes, and Manfred Tscheligi. The USUS evaluation framework for human-robot interaction. In *AISB2009: Proceedings of the Symposium on New Frontiers in Human-Robot Interaction*, volume 4, pages 11–26, 2009.

Tracey L Weissgerber, Natasa M Milic, Stacey J Winham, and Vesna D Garovic. Beyond bar and line graphs: time for a new data presentation paradigm. *PLoS Biology*, 13(4), 2015.

Tracey L Weissgerber, Marko Savic, Stacey J Winham, Dejana Stanisavljevic, Vesna D Garovic, and Natasa M Milic. Data visualization, bar naked: A free tool for creating interactive graphics. *Journal of Biological Chemistry*, 292(50):20592–20598, 2017.

Katherine S. Welfare, Matthew R. Hallowell, Julie A. Shah, and Laurel D. Riek. Consider the human work experience when integrating robotics in the workplace. In *2019 14th ACM/IEEE International Conference on Human-Robot Interaction (HRI)*, pages 75–84, 2019.

Yueh-Hsuan Weng, Yasuhisa Hirata, Osamu Sakura, and Yusuke Sugahara. The religious impacts of Taoism on ethically aligned design in HRI. *International Journal of Social Robotics*, 11(5):829–839, 2019. doi: 10.1007/s12369-019-00594-z.

Franz Werner. A survey on current practices in user evaluation of companion robots. In *Human-Robot Interaction*, pages 65–88. Springer, 2020.

Jelte M Wicherts, Coosje LS Veldkamp, Hilde EM Augusteijn, Marjan Bakker, Robbie CM Van Aert, and Marcel ALM Van Assen. Degrees of freedom in planning, running, analyzing, and reporting psychological studies: A checklist to avoid p-hacking. *Frontiers in Psychology*, 7:1832, 2016.

Noah Wiederhold, Ava Megyeri, DiMaggio Paris, Sean Banerjee, and Natasha Banerjee. HOH: Markerless multimodal human-object-human handover dataset with large object count. *Advances in Neural Information Processing Systems*, 36, 2024.

Rand R Wilcox and Guillaume A Rousselet. A guide to robust statistical methods in neuroscience. *Current Protocols in Neuroscience*, 82(1):8–42, 2018.

Tom Williams, Cynthia Matuszek, Ross Mead, and Nick Depalma. Scarecrows in Oz: the use of large language models in HRI. *ACM Transactions on Human-Robot Interaction*, 13(1):1–11, 2024.

Clavia Williams-McBean. The value of a qualitative pilot study in a multi-phase mixed methods research. *The Qualitative Report*, 24(5):1055–1064, 2019.

Thomas D Wilson and David R Streatfield. Structured observation in the investigation of information needs. *Social science information studies*, 1(3):173–184, 1981.

Katie Winkle, Erik Lagerstedt, Ilaria Torre, and Anna Offenwanger. 15 years of (who) man robot interaction: Reviewing the h in human-robot interaction. *ACM Transactions on Human-Robot Interaction*, 12(3):1–28, 2023.

Peter Wittenburg, Hennie Brugman, Albert Russel, Alex Klassmann, and Han Sloetjes. Elan: A professional framework for multimodality research. In *5th International Conference on Language Resources and Evaluation (LREC 2006)*, pages 1556–1559, 2006.

Lesley Wood, Matthias Egger, Lise Lotte Gluud, Kenneth F Schulz, Peter Jüni, Douglas G Altman, Christian Gluud, Richard M Martin, Anthony J G Wood, and Jonathan A C Sterne. Empirical evidence of bias in treatment effect estimates in controlled trials with different interventions and outcomes: meta-epidemiological study. *BMJ*, 336(7644):601, 2008. http://bmj.com/content/336/7644/601.full.pdf.

Sebastian Wrede, Christian Emmerich, Ricarda Grünberg, Arne Nordmann, Agnes Swadzba, and Jochen Steil. A user study on kinesthetic teaching of redundant robots in task and configuration space. *Journal of Human-Robot Interaction*, 2(1):56–81, 2013. doi: 10.5898/JHRI.2.1.Wrede.

Tina LY Wu, Anna Murphy, Chao Chen, and Dana Kulić. Human-in-the-loop auditory cueing strategy for gait modification. *IEEE Robotics and Automation Letters*, 6(2):3521–3528, 2021.

Ya-Huei Wu, Jérémy Wrobel, Mélanie Cornuet, Hélène Kerhervé, Souad Damnée, and Anne-Sophie Rigaud. Acceptance of an assistive robot in older adults: a mixed-method study of human–robot interaction over a 1-month period in the living lab setting. *Clinical Interventions in Aging*, 9:801, 2014.

Stephen Xia and Xiaofan Jiang. Pams: Improving privacy in audio-based mobile systems. In *Proceedings of the 2nd International Workshop on Challenges in Artificial Intelligence and Machine Learning for Internet of Things*, pages 41–47, 2020.

Yuki Yamashita, Hisashi Ishihara, Takashi Ikeda, and Minoru Asada. Path analysis for the halo effect of touch sensations of robots on their personality impressions. In *International Conference on Social Robotics*, pages 502–512. Springer, 2016.

H.A. Yanco and J. Drury. Classifying human-robot interaction: an updated taxonomy. In *2004 IEEE International Conference on Systems, Man and Cybernetics (IEEE Cat. No.04CH37583)*, volume 3, pages 2841–2846, 2004. doi: 10.1109/ICSMC.2004.1400763.

Holly A Yanco, Adam Norton, Willard Ober, David Shane, Anna Skinner, and Jack Vice. Analysis of human-robot interaction at the DARPA robotics challenge trials. *Journal of Field Robotics*, 2015.

Yan-Jun Yang, Soumith Udatha, Dana Kulić, and Elahe Abdi. A novel foot interface versus voice for controlling a robotic endoscope holder. In *2020 8th IEEE RAS/EMBS International Conference for Biomedical Robotics and Biomechatronics (BioRob)*, pages 272–279. IEEE, 2020.

Yan-Jun Yang, Arvind N Vadivelu, Charles HC Pilgrim, Dana Kulić, and Elahe Abdi. A novel perception framework for automatic laparoscope zoom factor control using tool geometry. In *2021 43rd Annual International Conference of the IEEE Engineering in Medicine & Biology Society (EMBC)*, pages 4700–4704. IEEE, 2021.

Yan-Jun Yang, Arvind Kumar N Vadivelu, Jessica Hepworth, Yongpeng Zeng, Charles HC Pilgrim, Dana Kulić, and Elahe Abdi. Experimental evaluation of accuracy and efficiency of two control strategies for a novel foot commanded robotic laparoscope holders with surgeons. *Scientific Reports*, 14(1):9264, 2024.

Payam Jome Yazdian, Mo Chen, and Angelica Lim. Gesture2Vec: Clustering gestures using representation learning methods for co-speech gesture generation. In *2022 IEEE/RSJ International Conference on Intelligent Robots and Systems (IROS)*, pages 3100–3107. IEEE, 2022.

Deng Yongda, Li Fang, and Xin Huang. Research on multimodal human-robot interaction based on speech and gesture. *Computers & Electrical Engineering*, 72:443–454, 2018. doi: 10.1016/j.compeleceng.2018.09.014.

James E Young, JaYoung Sung, Amy Voida, Ehud Sharlin, Takeo Igarashi, Henrik I Christensen, and Rebecca E Grinter. Evaluating human-robot interaction: Focusing on the holistic interaction experience. *International Journal of Social Robotics*, 3:53–67, 2011.

LAPARO Sp. z o.o. Laparo medical simulator training program. https://laparosimulators.com/laparo-training-program/, 2021. [Online; accessed 23-October-2021].

Oded Zafrani and Galit Nimrod. Towards a Holistic Approach to Studying Human-Robot Interaction in Later Life. *The Gerontologist*, 59(1):e26–e36, 2019. doi: 10.1093/geront/ gny077.

JD Zamfirescu-Pereira, David Sirkin, David Goedicke, Ray LC, Natalie Friedman, Ilan Mandel, Nikolas Martelaro, and Wendy Ju. Fake it to make it: Exploratory prototyping in HRI. In *Companion of the 2021 ACM/IEEE International Conference on Human-Robot Interaction*, pages 19–28, 2021a.

JD Zamfirescu-Pereira, David Sirkin, David Goedicke, Ray Lc, Natalie Friedman, Ilan Mandel, Nikolas Martelaro, and Wendy Ju. Fake it to make it: Exploratory prototyping in HRI. In *Companion of the 2021 ACM/IEEE International Conference on Human-Robot Interaction*, pages 19–28, 2021b.

Ahmed Zihni, William D Gerull, Jaime A Cavallo, Tianjia Ge, Shuddhadeb Ray, Jason Chiu, L Michael Brunt, and Michael M Awad. Comparison of precision and speed in laparoscopic and robot-assisted surgical task performance. *Journal of Surgical Research*, 223:29–33, 2018.

Siyang Zuo, Teng Chen, Xin Chen, and Baojun Chen. A wearable hands-free human-robot interface for robotized flexible endoscope. *IEEE Robotics and Automation Letters*, 7(2):3953–3960, 2022.

Index

D

E

J

K

S